Pauli Lectures on Physics

Volume 4
Statistical Mechanics

Wolfgang Pauli

edited by C. P. Enz

Foreword by
Victor F. Weisskopf

Pauli Lectures on Physics: Statistical Mechanics

Pauli Lectures on Physics

1. Electrodynamics
2. Optics and the Theory of Electrons
3. Thermodynamics and the Kinetic Theory of Gases
4. Statistical Mechanics
5. Wave Mechanics
6. Selected Topics in Field Quantization

Pauli Lectures on Physics: Volume 4. Statistical Mechanics

Wolfgang Pauli

Edited by Charles P. Enz

Translated by H. R. Lewis and S. Margulies

Foreword by Victor F. Weisskopf

The MIT Press
Cambridge, Massachusetts, and London, England

First MIT Press paperback edition, 1977
Second printing, 1978
Third printing, 1979
Fourth printing, 1981

ISBN 0 262 66035 0 (paperback)
Library of Congress catalog card number: 72-7804

Contents

Foreword

It is often said that scientific texts quickly become obsolete. Why are the Pauli lectures brought to the public today, when some of them were given as long as twenty years ago? The reason is simple: Pauli's way of presenting physics is never out of date. His famous article on the foundations of quantum mechanics appeared in 1933 in the German encyclopedia *Handbuch der Physik.* Twenty-five years later it reappeared practically unchanged in a new edition, whereas most other contributions to this encyclopedia had to be completely rewritten. The reason for this remarkable fact lies in Pauli's style, which is commensurate to the greatness of its subject in its clarity and impact. Style in scientific writing is a quality that today is on the point of vanishing. The pressure of fast publication is so great that people rush into print with hurriedly written papers and books that show little concern for careful formulation of ideas. Mathematical and instrumental techniques have become complicated and difficult; today most of the effort of writing and learning is devoted to the acquisition of these techniques instead of insight into important concepts. Essential ideas of physics are often lost in the dense forest of mathematical reasoning. This situation need not be so. Pauli's lectures show how physical ideas can be presented clearly and in good mathematical form, without being hidden in formalistic expertise.

Pauli was not an accomplished lecturer in the technical sense of the word. It was often difficult to follow his courses. But when the sequence of his thoughts and the structure of his logic become apparent, the attentive follower is left with a new and deeper knowledge of essential concepts and with a clearer insight into the splendid architecture of reason, which is theoretical physics. The value of the lecture notes is not diminished by the fact that they were written not by him but by some of his collaborators. They bear the mark of the master in their conceptual structure and their mathematical rigidity. Only here and there does one miss words and comments of the master. Neither does one notice the passing of time in his lectures, with the sole exception of the lectures on field quantization, in which some concepts are formulated in a way that may appear old-fashioned to some today. But even these lectures should be of use to modern students because of their compactness and their direct approach to the central problems.

May this volume serve as an example of how the concepts of theoretical physics were conceived and taught by one of the great men who created them.

Victor F. Weisskopf

Cambridge, Massachusetts

Preface

"Statistische Mechanik" was the first of the six volumes of lecture notes of Pauli published by the "Verein der Mathematiker und Physiker an der E.T.H. Zürich." After many years of lecturing, Pauli's personal notes had started to disintegrate. So he asked his assistant, who at that time was the late M. R. Schafroth, to take notes for his personal use. The result was a short compendium with almost no text. But the students asked Pauli for the permission to publish the notes for their use, which they did in 1947. It was indeed far from easy for an uninitiated student to follow a course given by Pauli.

But Schafroth's notes did not only reflect the original purpose of a compendium for Pauli's personal use. They were also typical of Schafroth's style of talking physics. It is the same style as that of the *Selected Topics in Field Quantization*, in this series, which was also based on notes by Schafroth.

These particular circumstances of the origin of the present lectures of course posed many problems both to the translators and to the editor. Phrases had to be inserted and some derivations needed changes and amplifications. The result, I believe, is a concise course on statistical mechanics giving today's student a quick, if not easy, introduction to the subject, centered on the historic development of the basic ideas and on the logical structure of the theory.

Since these lectures were given by Pauli important advances have been made in the rigorous solution of simple models of statistical mechanics which have given detailed insight into the phenomenon of phase transitions. The first and most spectacular achievement in this field, Onsager's solution of the two-dimensional Ising model, although published three years before the German original of these lectures, had not been discussed by Pauli. Most likely at that time Pauli considered this problem too advanced and technical a subject for the level of this course. And, in fact, since then the technique of deriving Onsager's result has been considerably simplified. A good introduction to this field and, more generally, to the problem of phase transitions can be found in the new book by H. E. Stanley, *Phase Transitions and Critical Phenomena* (Oxford University Press, New York, 1971).

Charles P. Enz

Geneva, 19 November 1971

Pauli Lectures on Physics:
Statistical Mechanics

Chapter 1. Stosszahlansatz[1]

1. CONCEPTS FROM THE ELEMENTARY KINETIC THEORY OF GASES

Let $f(\boldsymbol{v})\,d^3v$ be the number of molecules with velocity contained in $d^3\boldsymbol{v}$, and let $n=\int f(\boldsymbol{v})\,d^3v$ be the number of molecules per cm^3. The function $f(\boldsymbol{v})$ can also depend on $\boldsymbol{x}$. Then, we have

$$\iint f(\boldsymbol{v},\boldsymbol{x})\,d^3v\,d^3x=\int n(\boldsymbol{x})\,d^3x=N\,,$$

where N is the total number of molecules.

The *pressure tensor* is defined by

$$p_{ik}=m\int v_i v_k f\,d^3v\,,\qquad p_{ik}=p_{ki}=nm\overline{v_i v_k}\,,$$

where m is the mass of the molecules. The *mean velocity* is defined by

$$c_i\equiv\bar{v}_i=\frac{1}{n}\int v_i f\,d^3v\,,$$

$$\boldsymbol{c}=\overline{\boldsymbol{v}}=\frac{1}{n}\int \boldsymbol{v} f\,d^3v\,.$$

If there is molecular flow, $\boldsymbol{c}\neq 0$, we define

$$\boldsymbol{u}=\boldsymbol{v}-\boldsymbol{c}\,.$$

[1] The "Stosszahlansatz" is an assumption related to the calculation of the number of collisions between the molecules of a gas.

Then, strictly speaking, the pressure tensor is defined as

$$p_{ik} = nm\overline{u_i u_k} = m\int u_i u_k f(\boldsymbol{u}+\boldsymbol{c})\,\mathrm{d}^3u\,.$$

Since $\bar{\boldsymbol{u}}=0$, we have

$$\overline{v_i v_k} = c_i c_k + \overline{u_i u_k}\,.$$

Example: Consider a homogeneous, isotropic velocity distribution:

$$c_i = 0\,,\qquad f(\boldsymbol{v}) = F(v)\,,\qquad v = |\boldsymbol{v}|\,,$$

$$p_{ik} = p\delta_{ik}\,,\qquad p = nm\overline{v_1^2} = nm\overline{v_2^2} = nm\overline{v_3^2}\,.$$

Thus,

$$p = \tfrac{1}{3}nm\overline{v^2}\,.$$

The ideal gas laws follow from this equation: Since $\frac{1}{2}nm\overline{v^2} = U =$ average kinetic energy of the molecules per cm³, the equation can be rewritten as

$$p = \tfrac{2}{3}U\,.$$

2. COLLISION LAWS

We consider collisions between two masses m and M which move with velocities $\boldsymbol{v}$ and $\boldsymbol{V}$, respectively. We define the relative velocity by

$$\boldsymbol{w} = \boldsymbol{v} - \boldsymbol{V}$$

and the velocity of the center of mass by

$$\boldsymbol{U} = \frac{m\boldsymbol{v} + M\boldsymbol{V}}{m+M}\,.$$

Solving for $\boldsymbol{v}$ and $\boldsymbol{V}$, we obtain

$$\left.\begin{aligned} \boldsymbol{v} &= \boldsymbol{U} + \frac{M}{M+m}\boldsymbol{w} \\ \boldsymbol{V} &= \boldsymbol{U} - \frac{m}{M+m}\boldsymbol{w} \end{aligned}\right\}. \qquad [2.1]$$

These equations, combined with the definition of the kinetic energy, $E_{\text{kin}} = \frac{1}{2}m\boldsymbol{v}^2 + \frac{1}{2}M\boldsymbol{V}^2$, give

$$E_{\text{kin}} = \frac{m+M}{2}\boldsymbol{U}^2 + \frac{1}{2}\frac{mM}{M+m}\boldsymbol{w}^2 \,. \qquad [2.2]$$

If primes are used to denote quantities which refer to times after the collision, then the conservation laws of momentum and energy may be written as

$$\left.\begin{aligned} &\boldsymbol{P}' = \boldsymbol{P} \rightarrow \boldsymbol{U}' = \boldsymbol{U} \\ &E'_{\text{kin}} = E_{\text{kin}} \end{aligned}\right\} . \qquad [2.3]$$

Substituting $\boldsymbol{U}' = \boldsymbol{U}$ into $E'_{\text{kin}} = E_{\text{kin}}$ gives

$$\boldsymbol{w}'^2 = \boldsymbol{w}^2 \,, \qquad |\boldsymbol{w}'| = |\boldsymbol{w}| \,. \qquad [2.4]$$

The conservation laws do not suffice for determining the variables after the collision. There are two pieces of information still lacking, which we can determine by means of special models.

a. Billiard ball model

We idealize the molecules by considering them to be rigid, elastic spheres. The collision is specified by a unit vector $\boldsymbol{n}$ which points from the center of m toward the

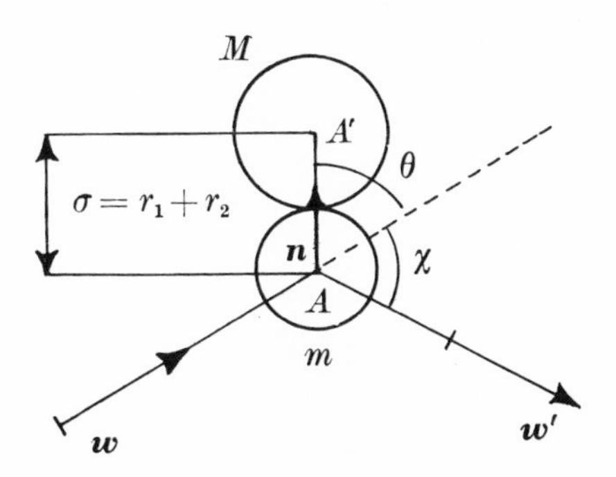

Figure 2.1

center of $\boldsymbol{M}$ at the time of collision. We then have

$$\boldsymbol{w}_n' = -\boldsymbol{w}_n\,, \qquad \boldsymbol{w}_{\perp n}' = \boldsymbol{w}_{\perp n}\,. \qquad [2.5]$$

In place of $\boldsymbol{n}$ we can introduce two angles: $\theta=$ angle between $\boldsymbol{w}$ and $\boldsymbol{n}$, and $\varepsilon=$ angle between the plane determined by $\boldsymbol{w}$ and $\boldsymbol{n}$ and a fixed plane containing $\boldsymbol{w}$. Then, the scattering angle is given by $\chi=\pi-2\theta$. If we vary the initial conditions a little, so that $\boldsymbol{U}$ is contained in d^3U and $\boldsymbol{w}$ is contained in d^3w, then $\boldsymbol{U}'$ will be contained in d^3U' and $\boldsymbol{w}'$ will be contained in d^3w'. The differentials are related by

$$d^3U = d^3U' \qquad \text{(from Eq. [2.3])}$$

$$d^3w = d^3w' \qquad \text{(from Eq. [2.5])}\,.$$

Therefore,

$$d^3w\, d^3U = d^3w'\, d^3U'\,. \qquad [2.6]$$

Furthermore, because of Eq. [2.1], we have

$$\frac{\partial(\boldsymbol{U}, \boldsymbol{w})}{\partial(\boldsymbol{V}, \boldsymbol{v})} = \prod_{i=1}^{3} \frac{\partial(U_i, w_i)}{\partial(V_i, v_i)} = 1\times1\times1 = 1\,. \qquad [2.7]$$

Using Eqs. [2.6] and [2.7], we obtain

$$d^3V\, d^3v = d^3V'\, d^3v'\,. \qquad [2.8]$$

If we allow $\boldsymbol{n}$ to vary within the solid angle $d^2\lambda = \sin\theta\, d\theta\, d\varepsilon$, then, because there is no $d^2\lambda'$, we have

$$d^3V\, d^3v\, d^2\lambda = d^3V'\, d^3v'\, d^2\lambda\,. \qquad [2.9]$$

This is a collision invariant which we shall find useful later.

b. General central force model

We assume that the molecules are mass points which repel each other according to some central force law. We can draw a diagram analogous to that of Fig. 2.1, in which the straight lines along $\boldsymbol{w}$ and $\boldsymbol{w}'$, instead of being the trajectory of the center of mass, are the asymptotes to the trajectory.

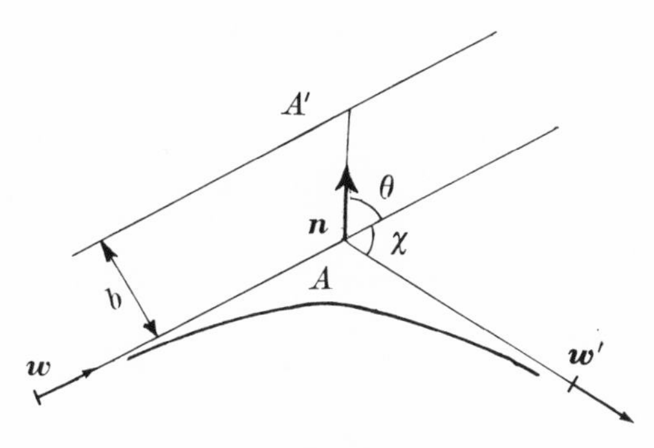

Figure 2.2

Then we can introduce the unit vector $\boldsymbol{n}$ which is parallel to the bisector of the angle between $\boldsymbol{w}$ and $\boldsymbol{w}'$ in the plane of the trajectory, and which points in the direction from m toward M. We can also introduce the impact parameter b, the scattering angle $\chi=\chi(|\boldsymbol{w}|, b)$, and the angle of the plane of the trajectory ε. As in Fig. 2.1, we define σ by $\sigma=b/\sin\theta$, where $\theta=(\pi-\chi)/2$. Then Eq. [2.5] still holds; therefore, Eqs. [2.6], [2.8], and [2.9] are also valid, and $d^3v\,d^3V$ is a collision invariant also in this case.

We introduce two additional concepts: the *inverse collision* and the *opposite collision.* They are characterized by the parameters in the table below.

Original Collision	*Inverse Collision*	*Opposite Collision*	
$\boldsymbol{v}, \boldsymbol{V};\ \theta, \varepsilon$	$-\boldsymbol{v}', -\boldsymbol{V}';\ \theta, \varepsilon$	$\boldsymbol{v}', \boldsymbol{V}';\ \theta, \pi+\varepsilon$	initial
$\boldsymbol{U}, \boldsymbol{w};\ \boldsymbol{n}$	$-\boldsymbol{U}', -\boldsymbol{w}';\ \boldsymbol{n}$	$\boldsymbol{U}', \boldsymbol{w}';\ -\boldsymbol{n}$	
$\boldsymbol{v}', \boldsymbol{V}'$	$-\boldsymbol{v}, -\boldsymbol{V}$	$\boldsymbol{v}, \boldsymbol{V}$	final
$\boldsymbol{U}', \boldsymbol{w}'$	$-\boldsymbol{U}, -\boldsymbol{w}$	$\boldsymbol{U}, \boldsymbol{w}$	

3. CHANGE OF THE DISTRIBUTION FUNCTION DUE TO COLLISIONS

We assume that there are two types of molecules: type 1 is characterized by f and $\boldsymbol{v}$, and type 2 is characterized

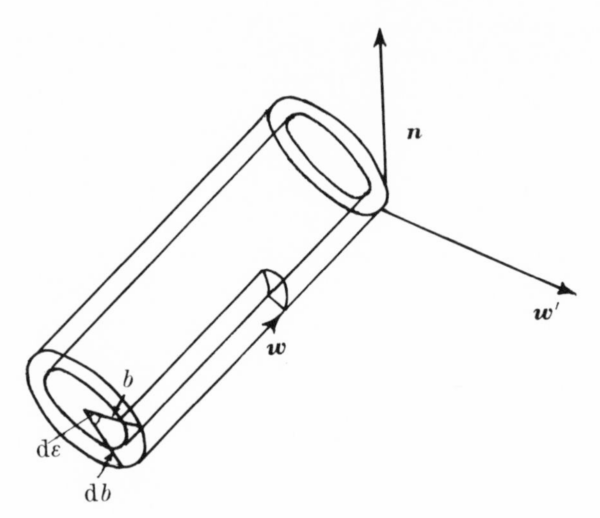

Figure 3.1

by $\boldsymbol{F}$ and $\boldsymbol{V}$. A decrease in the number of molecules of type 1 in $\mathrm{d}^3\boldsymbol{v}$ by collisions arises from *a*) collisions with molecules of type 2, and *b*) collisions with molecules of type 1. The number of collisions per second, characterized by $\boldsymbol{b}$ and ε, between molecules of type 2 whose velocities are in d^3V and molecules of type 1 whose velocities are in $\mathrm{d}^3\boldsymbol{v}$ is

$$w b \,\mathrm{d}b \,\mathrm{d}\varepsilon f(\boldsymbol{v}) F(\boldsymbol{V}) \,\mathrm{d}^3v \,\mathrm{d}^3V \quad .$$

(see Fig. 3.1). This implies the following (negative) rate of change of the number of molecules of type 1 in $\mathrm{d}^3\boldsymbol{v}$:

$$\mathrm{d}^3v \frac{\partial f}{\partial t} = -\iiint \mathrm{d}^3V \,\mathrm{d}b \,\mathrm{d}\varepsilon \, w b f(\boldsymbol{v}) F(\boldsymbol{V}) \,\mathrm{d}^3v \,.$$

The rate of change due to the decrease of the number of molecules in d^3v because of collisions between like molecules (type 1) is

$$\mathrm{d}^3v \frac{\partial f}{\partial t} = -\iiint \mathrm{d}^3V \,\mathrm{d}b \,\mathrm{d}\varepsilon \, w b f(\boldsymbol{v}) f(\boldsymbol{V}) \,\mathrm{d}^3v \,.$$

Similar consideration of the increase of the number of molecules in d^3v due to collisions leads to

$$\mathrm{d}^3v\,\frac{\partial f}{\partial t}=+\iiint \mathrm{d}^3V'\,\mathrm{d}b\,\mathrm{d}\varepsilon\, wb\, f(\boldsymbol{v}')F(\boldsymbol{V}')\,\mathrm{d}^3v' + \iiint \mathrm{d}^3V'\,\mathrm{d}b\,\mathrm{d}\varepsilon\, wb\, f(\boldsymbol{v}')f(\boldsymbol{V}')\,\mathrm{d}^3v' .$$

In this expression, the given quantities are $\boldsymbol{v}$ and $\boldsymbol{V}$ rather than $\boldsymbol{v}'$ and $\boldsymbol{V}'$. It would be more precise to write

$$\boldsymbol{v}'=\boldsymbol{v}'(\boldsymbol{v},\boldsymbol{V},b,\varepsilon) \qquad \text{and} \qquad \boldsymbol{V}'=\boldsymbol{V}'(\boldsymbol{v},\boldsymbol{V},b,\varepsilon) .$$

By application of Eq. [2.8] it follows that

$$-\frac{\partial f}{\partial t}=\iiint wb\,\mathrm{d}b\,\mathrm{d}\varepsilon\,\mathrm{d}^3V[f(\boldsymbol{v})F(\boldsymbol{V})-f(\boldsymbol{v}')F(\boldsymbol{V}')] + \iiint wb\,\mathrm{d}b\,\mathrm{d}\varepsilon\,\mathrm{d}^3V[f(\boldsymbol{v})f(\boldsymbol{V})-f(\boldsymbol{v}')f(\boldsymbol{V}')] .$$

The impact parameter is a function of θ and w, $b=b(\theta,w)$. Therefore,

$$b\,\mathrm{d}b\,\mathrm{d}\varepsilon=b\left(\frac{\partial b}{\partial \theta}\right)_w \mathrm{d}\theta\,\mathrm{d}\varepsilon=Q\,\mathrm{d}^2\lambda ,$$

where

$$Q=b\left(\frac{\partial b}{\partial \theta}\right)_w \frac{1}{\sin\theta} .$$

For rigid spheres, $Q=\sigma^2\cos\theta$. (In what follows, we use Q for M-M collisions, $\bar{Q}$ for M-m collisions, and q for m-m collisions.)

Finally, we have

$$-\frac{\partial f(\boldsymbol{v})}{\partial t}=\iint w\bar{Q}\,\mathrm{d}^2\lambda\,\mathrm{d}^3V[f(\boldsymbol{v})F(\boldsymbol{V})-f(\boldsymbol{v}')F(\boldsymbol{V}')] + \iint wq\,\mathrm{d}^2\lambda\,\mathrm{d}^3V[f(\boldsymbol{v})f(\boldsymbol{V})-f(\boldsymbol{v}')f(\boldsymbol{V}')] ,$$

and

$$-\frac{\partial F(\boldsymbol{V})}{\partial t}=\iint w\bar{Q}\,\mathrm{d}^2\lambda\,\mathrm{d}^3v[f(\boldsymbol{v})F(\boldsymbol{V})-f(\boldsymbol{v}')F(\boldsymbol{V}')] + \iint wQ\,\mathrm{d}^2\lambda\,\mathrm{d}^3v[F(\boldsymbol{v})F(\boldsymbol{V})-F(\boldsymbol{v}')F(\boldsymbol{V}')] .$$

Remark: In deriving this equation, we have introduced a fundamental hypothesis which is in addition to the assumption of central forces only; this hypothesis is called the "Stosszahlansatz." We have assumed that the density of molecules in the cylindrical volume element $b\,\mathrm{d}b\,\mathrm{d}\varepsilon$ is the same as it is in the remaining gas. Since, however, there will be density fluctuations, $\partial f/\partial t$ and $\partial F/\partial t$ will also fluctuate. By neglecting these fluctuations, we actually have not calculated $\partial f/\partial t$; instead, we have calculated $\Delta f/\Delta t$. This one-sided average also means that our formula picks out a specific direction in time; this is in contrast to the laws which were used in the derivation.

4. STATIONARY DISTRIBUTIONS

Case 1. *Distribution functions independent of position*

We define the Boltzmann H-function,

$$H_B \equiv \int f \log f \,\mathrm{d}^3 v + \int F \log F \,\mathrm{d}^3 V, \tag{4.1}$$

and a related function,

$$H \equiv \int f(\log f - 1)\,\mathrm{d}^3 v + \int F(\log F - 1)\,\mathrm{d}^3 V = H_B - n_m - n_M \,. \tag{4.2}$$

Because of number conservation, it follows that

$$\begin{aligned}\frac{\mathrm{d}H}{\mathrm{d}t} &= \frac{\mathrm{d}H_B}{\mathrm{d}t} \\ &= \int \log f \frac{\partial f}{\partial t}\,\mathrm{d}^3 v + \int \log F \frac{\partial F}{\partial t}\,\mathrm{d}^3 V \\ &= -\iiint w\bar{Q}\,\mathrm{d}^2\lambda\,\mathrm{d}^3 v\,\mathrm{d}^3 V[\log f(\boldsymbol{v}) + \log F(\boldsymbol{V})] \\ &\qquad\qquad \times [f(\boldsymbol{v})F(\boldsymbol{V}) - f(\boldsymbol{v}')F(\boldsymbol{V}')] \\ &\quad -\iiint wq\,\mathrm{d}^2\lambda\,\mathrm{d}^3 v\,\mathrm{d}^3 V[\log f(\boldsymbol{v})][f(\boldsymbol{v})f(\boldsymbol{V}) - f(\boldsymbol{v}')f(\boldsymbol{V}')] \\ &\quad -\iiint wQ\,\mathrm{d}^2\lambda\,\mathrm{d}^3 v\,\mathrm{d}^3 V[\log F(\boldsymbol{V})][F(\boldsymbol{v})F(\boldsymbol{V}) - F(\boldsymbol{v}')F(\boldsymbol{V}')] \,.\end{aligned}$$

We have

(*a*) $$\iiint wq\,\mathrm{d}^2\lambda\,\mathrm{d}^3v\,\mathrm{d}^3V[\log f(\boldsymbol{v})]f(\boldsymbol{v})\,f(\boldsymbol{V})$$
$$=\frac{1}{2}\iiint wq\,\mathrm{d}^2\lambda\,\mathrm{d}^3v\,\mathrm{d}^3V[\log f(\boldsymbol{v})+\log f(\boldsymbol{V})]f(\boldsymbol{v})\,f(\boldsymbol{V})$$
$$=\frac{1}{2}\iiint wq\,\mathrm{d}^2\lambda\,\mathrm{d}^3v\,\mathrm{d}^3V\{\log[f(\boldsymbol{v})f(\boldsymbol{V})]\}\,f(\boldsymbol{v})f(\boldsymbol{V})\,.$$

(*b*) Because the molecules are identical, interchanging $\boldsymbol{v}$ and $\boldsymbol{V}$ also interchanges $\boldsymbol{v}'$ and $\boldsymbol{V}'$:

$$\iiint wq\,\mathrm{d}^2\lambda\,\mathrm{d}^3v\,\mathrm{d}^3V[\log f(\boldsymbol{v})]f(\boldsymbol{v}')f(\boldsymbol{V}')$$
$$=\frac{1}{2}\iiint \omega q\,\mathrm{d}^2\lambda\,\mathrm{d}^3v\,\mathrm{d}^3V\{\log[f(\boldsymbol{v})f(\boldsymbol{V})]\}\,f(\boldsymbol{v}')f(\boldsymbol{V}')\,.$$

(*c*) With these results, we obtain for the second integral in the expression for $\mathrm{d}H/\mathrm{d}t$, for example,

$$-\frac{1}{2}\iiint wq\,\mathrm{d}^2\lambda\,\mathrm{d}^3v\,\mathrm{d}^3V\{\log[f(\boldsymbol{v})f(\boldsymbol{V})]\}\{f(\boldsymbol{v})f(\boldsymbol{V})-f(\boldsymbol{v}')f(\boldsymbol{V}')\}\,.$$

(*d*) Because of the properties of opposite collisions, this integral can be written as

$$-\frac{1}{4}\iiint wq\,\mathrm{d}^2\lambda\,\mathrm{d}^3v\,\mathrm{d}^3V\{\log[f(\boldsymbol{v})f(\boldsymbol{V})]-\log[f(\boldsymbol{v}')f(\boldsymbol{V}')]\}$$
$$\times\{f(\boldsymbol{v})f(\boldsymbol{V})-f(\boldsymbol{v}')f(\boldsymbol{V}')\}\,.$$

Finally, it follows that

$$-\frac{\mathrm{d}H}{\mathrm{d}t}=\frac{1}{2}\iiint w\bar{Q}\,\mathrm{d}^2\lambda\,\mathrm{d}^3v\,\mathrm{d}^3V\{\log[f(\boldsymbol{v})F(\boldsymbol{V})]-\log[f(\boldsymbol{v}')F(\boldsymbol{V}')]\}$$
$$\times[f(\boldsymbol{v})F(\boldsymbol{V})-f(\boldsymbol{v}')F(\boldsymbol{V}')]$$
$$+\frac{1}{4}\iiint wq\,\mathrm{d}^2\lambda\,\mathrm{d}^3v\,\mathrm{d}^3V\{\log[f(\boldsymbol{v})f(\boldsymbol{V})]-\log[f(\boldsymbol{v}')f(\boldsymbol{V}')]\}$$
$$\times\{f(\boldsymbol{v})f(\boldsymbol{V})-f(\boldsymbol{v}')f(\boldsymbol{V}')\}$$
$$+\frac{1}{4}\iiint wQ\,\mathrm{d}^2\lambda\,\mathrm{d}^3v\,\mathrm{d}^3V\{\log[F(\boldsymbol{v})F(\boldsymbol{V})]-\log[F(\boldsymbol{v}')F(\boldsymbol{V}')]\}$$
$$\times\{F(\boldsymbol{v})F(\boldsymbol{V})-F(\boldsymbol{v}')F(\boldsymbol{V}')\}\,.$$

Since
$$(x-y)(\log x-\log y)>0\,,\qquad \text{for } x\neq y\,,$$
$$=0\,,\qquad \text{for } x=y\,,$$

we therefore obtain the Boltzmann H-theorem:

$$\frac{\mathrm{d}H}{\mathrm{d}t}\leqslant 0\,. \qquad [4.3]$$

We are seeking stationary conditions, that is, distributions f and F for which $\partial f/\partial t=0$ and $\partial F/\partial t=0$. From this it follows that $\mathrm{d}H/\mathrm{d}t=0$, which is therefore a necessary condition for a stationary distribution. Because $\mathrm{d}H/\mathrm{d}t=0$, we have

$$f(\boldsymbol{v})\;f(\boldsymbol{V})=\;f(\boldsymbol{v}')\;f(\boldsymbol{V}') \qquad [4.4]$$
$$F(\boldsymbol{v})F(\boldsymbol{V})=F(\boldsymbol{v}')F(\boldsymbol{V}') \qquad [4.5]$$
$$f(\boldsymbol{v})F(\boldsymbol{V})=\;f(\boldsymbol{v}')F(\boldsymbol{V}')\,. \qquad [4.6]$$

These relations must hold for all $\boldsymbol{v}'$ and $\boldsymbol{V}'$ which are allowed for the values of $\boldsymbol{v}$ and $\boldsymbol{V}$; that is, they must hold for all $\boldsymbol{v}'$ and $\boldsymbol{V}'$ which satisfy the energy and momentum conservation laws:

$$\boldsymbol{p}=\;m\boldsymbol{v}+m\boldsymbol{V}=m\boldsymbol{v}'+m\boldsymbol{V}'=\boldsymbol{p}'\,, \qquad [4.7a]$$
$$\boldsymbol{P}=M\boldsymbol{v}+M\boldsymbol{V}=M\boldsymbol{v}'+M\boldsymbol{V}'=\boldsymbol{P}'\,, \qquad [4.8a]$$
$$\overline{\boldsymbol{P}}=\;m\boldsymbol{v}+M\boldsymbol{V}=\;m\boldsymbol{v}'+M\boldsymbol{V}'=\overline{\boldsymbol{P}}'\,, \qquad [4.9a]$$

$$e=\frac{m}{2}v^2+\frac{m}{2}V^2=\frac{m}{2}v'^2+\frac{m}{2}V'^2=e'\,, \qquad [4.7b]$$

$$E=\frac{M}{2}v^2+\frac{M}{2}V^2=\frac{M}{2}v'^2+\frac{M}{2}V'^2=E'\,, \qquad [4.8b]$$

$$\bar{E}=\frac{m}{2}v^2+\frac{M}{2}V^2=\frac{m}{2}v'^2+\frac{M}{2}V'^2=\bar{E}'\,. \qquad [4.9b]$$

However, from this it follows that $\partial f/\partial t=0$ and $\partial F/\partial t=0$; that is, $\mathrm{d}H/\mathrm{d}t=0$ is also a sufficient condition for a stationary distribution. If we substitute

$$\log f=\varphi \qquad \text{and} \qquad \log F=\Phi\,,$$

then Eqs. [4.4]–[4.6] become

$$\varphi(\boldsymbol{v}) + \varphi(\boldsymbol{V}) = \varphi(\boldsymbol{v}') + \varphi(\boldsymbol{V}') , \qquad [4.4a]$$

$$\Phi(\boldsymbol{v}) + \Phi(\boldsymbol{V}) = \Phi(\boldsymbol{v}') + \Phi(\boldsymbol{V}') , \qquad [4.5a]$$

and

$$\varphi(\boldsymbol{v}) + \Phi(\boldsymbol{V}) = \varphi(\boldsymbol{v}') + \Phi(\boldsymbol{V}') , \qquad [4.6a]$$

with Eqs. [4.7a]–[4.9b] as auxiliary conditions.

If we treat the problem with Lagrange multipliers, we obtain

$$\left[\frac{\partial\varphi}{\partial\boldsymbol{v}} + \alpha' m\boldsymbol{v} + \boldsymbol{\beta} m\right]\cdot \mathrm{d}\boldsymbol{v} + [\ldots]\cdot \mathrm{d}\boldsymbol{V} = [\ldots]\cdot \mathrm{d}\boldsymbol{v}' + [\ldots]\cdot \mathrm{d}\boldsymbol{V}'. \qquad [4.4b]$$

The quantity in each bracket must equal zero:

$$\frac{\partial\varphi}{\partial\boldsymbol{v}} + \alpha' m\boldsymbol{v} + \boldsymbol{\beta} m = 0 .$$

The solution is

$$f = A' \exp[-\alpha' \tfrac{1}{2} m\boldsymbol{v}^2 - \boldsymbol{\beta}\cdot m\boldsymbol{v}] .$$

If we define $\boldsymbol{c} = -\boldsymbol{\beta}/\alpha'$ and $\alpha' m/2 = \alpha$, $A' \exp(\alpha c^2) = A$, then

$$f = A \exp[-\alpha(\boldsymbol{v} - \boldsymbol{c})^2] .$$

This is a Maxwell distribution with a superimposed constant molecular flow $\boldsymbol{c}$.

Similarly, from Eq. [4.5a] we obtain

$$F = \bar{A}' \exp[-a' \tfrac{1}{2} M\boldsymbol{V}^2 - \boldsymbol{b}\cdot M\boldsymbol{V}] .$$

Since, from Eq. [4.6], fF can only depend on $\bar{E}$ and $\bar{\boldsymbol{P}}$, it follows that $a' = \alpha'$ and $\boldsymbol{b} = \boldsymbol{\beta}$, which means that $\boldsymbol{c}$ is the same in both cases. Therefore, the two gases may not move relative to one another. Thus, it follows that the most general stationary distribution independent of position is the Maxwell distribution:

$$f = A \exp[-\alpha(\boldsymbol{v} - \boldsymbol{c})^2] ,$$

$$F = \bar{A} \exp[-a(\boldsymbol{v} - \boldsymbol{c})^2] .$$

Case 2. Dependence of f on position and effects of external forces

These two factors give rise to additional terms in the expression for $\partial f/\partial t$.

(*a*) *Dependence on position.* From

$$f(t+\mathrm{d}t,\, \boldsymbol{x},\, \boldsymbol{v}) = f(t,\, \boldsymbol{x} - \boldsymbol{v}\,\mathrm{d}t,\, \boldsymbol{v})$$

follows

$$\frac{\partial f}{\partial t}\mathrm{d}t = -\frac{\partial f}{\partial \boldsymbol{x}}\cdot\boldsymbol{v}\,\mathrm{d}t\,.$$

Thus, the contribution to the time derivative of f due to motion of the molecules is

$$-\frac{\partial f}{\partial \boldsymbol{x}}\cdot\boldsymbol{v}\,.$$

(*b*) *Effect of external forces.* In this case

$$m\,\frac{\mathrm{d}\boldsymbol{v}}{\mathrm{d}t} = \boldsymbol{K}\,;$$

hence

$$f(t+\mathrm{d}t,\, \boldsymbol{x},\, \boldsymbol{v}) = f\left(t,\, \boldsymbol{x},\, \boldsymbol{v} - \frac{\boldsymbol{K}}{m}\mathrm{d}t\right).$$

Thus, the contribution to the time derivative is

$$-\frac{\partial f}{\partial \boldsymbol{v}}\cdot\frac{\boldsymbol{K}}{m}\,.$$

With these results we obtain the complete basic equation of the kinetic theory of a one-component gas [A-1] [2]:

$$\frac{\partial f}{\partial t} + \frac{\partial f}{\partial \boldsymbol{x}}\cdot\boldsymbol{v} + \frac{\partial f}{\partial \boldsymbol{v}}\cdot\frac{\boldsymbol{K}}{m} = \iint \mathrm{d}^3V\,\mathrm{d}^2\lambda[f(\boldsymbol{v}')f(\boldsymbol{V}') - f(\boldsymbol{v})f(\boldsymbol{V})]wq\,.$$

[2] Comments [A-1]–[A-8] appear in the Appendix on pp. 111–113.

The generalization to two- and more-component gases is immediate.

In order to find stationary distributions, let us assume that $\boldsymbol{K}$ is a conservative field:

$$\boldsymbol{K} = -\frac{\partial E_{\text{pot}}}{\partial \boldsymbol{x}}.$$

From the assumption that the distribution is stationary, it follows that $\partial f/\partial t = 0$ and that the change of f by collisions must vanish; that is, the right-hand side must vanish. Therefore,

(1) f is a Maxwell distribution with position-dependent constants, and

$$m\boldsymbol{v} \cdot \frac{\partial f}{\partial \boldsymbol{x}} - \frac{\partial E_{\text{pot}}}{\partial \boldsymbol{x}} \cdot \frac{\partial f}{\partial \boldsymbol{v}} = 0. \tag{2}$$

This is satisfied if f depends only on E_{kin} and E_{pot}:

$$\frac{\partial f}{\partial \boldsymbol{x}} = \frac{\partial f}{\partial E} \frac{\partial E_{\text{pot}}}{\partial \boldsymbol{x}},$$

$$\frac{\partial f}{\partial \boldsymbol{v}} = \frac{\partial f}{\partial E} m\boldsymbol{v};$$

hence

$$\frac{\partial f}{\partial \boldsymbol{x}} \cdot m\boldsymbol{v} - \frac{\partial E_{\text{pot}}}{\partial \boldsymbol{x}} \cdot \frac{\partial f}{\partial \boldsymbol{v}} = 0.$$

Thus, we obtain a stationary distribution if we replace $E_{\text{kin}} = \frac{1}{2}mv^2$ by $E = E_{\text{kin}} + E_{\text{pot}}$ in the Maxwell distribution. The resulting distribution is called the Maxwell-Boltzmann distribution:

$$f(\boldsymbol{v}, \boldsymbol{x}) = A' \exp\left[-\alpha' \left(\frac{m}{2} v^2 + E_{\text{pot}}(x)\right)\right].$$

Remark concerning the exact derivation of this formula. We must begin with a generalized H-theorem in which the position-dependent H-function is integrated over the gas volume G:

$$\mathscr{H} = \int_G \mathrm{d}^3x\, H(\boldsymbol{x}) = \int_G \mathrm{d}^3x \int_\infty \mathrm{d}^3v\, (f \log f - f)\,,$$

$$\frac{\mathrm{d}\mathscr{H}}{\mathrm{d}t} = \int_\infty \mathrm{d}^3v \int_G \mathrm{d}^3x \log f \frac{\partial f}{\partial t}$$

$$= J - \int_\infty \mathrm{d}^3v \int_G \mathrm{d}^3x \log f \frac{\partial f}{\partial \boldsymbol{x}} \cdot \boldsymbol{v} - \int_\infty \mathrm{d}^3v \int_G \mathrm{d}^3x \log f \frac{\partial f}{\partial \boldsymbol{v}} \cdot \frac{\boldsymbol{K}}{m}$$

$$= J - \int_\infty \mathrm{d}^3v \int_\Sigma \mathrm{d}\boldsymbol{\sigma} \cdot \boldsymbol{v}\, (f \log f - f) - \int_G \mathrm{d}^3x \int_\Omega \mathrm{d}\boldsymbol{\omega} \cdot \frac{\boldsymbol{K}}{m} (f \log f - f)\,.$$

Here, $J \leqslant 0$ and Σ is the boundary of G. The integral over the infinite sphere Ω in $\boldsymbol{v}$-space clearly vanishes.

The second integral can be made to vanish if we assume that Σ consists of ideally reflecting walls, or that the gas is not bounded, so that Σ can be placed at infinity. Then,

$$\frac{\mathrm{d}\mathscr{H}}{\mathrm{d}t} = J = \int_G \mathrm{d}^3x \int \mathrm{d}^3v \int \mathrm{d}^3V \int \mathrm{d}^2\lambda$$

$$\times [f(\boldsymbol{v}')f(\boldsymbol{V}') - f(\boldsymbol{v})f(\boldsymbol{V})]qw \log f \leqslant 0\,.$$

For stationary distributions we have $\partial f/\partial t = 0$, which implies $\mathrm{d}\mathscr{H}/\mathrm{d}t = 0$. Again, this is only possible for

$$f(\boldsymbol{v}')f(\boldsymbol{V}') = f(\boldsymbol{v})f(\boldsymbol{V})$$

with specified auxiliary conditions. The result is

$$f = A \exp[-\alpha(\boldsymbol{v} - \boldsymbol{c})^2]\,,$$

where A, α, and $\boldsymbol{c}$ are functions of $\boldsymbol{x}$. This satisfies the basic equation and the condition for a stationary distribution, $\partial f/\partial t = 0$, only if [A-1]

$$\frac{\partial f}{\partial \boldsymbol{x}} \cdot \boldsymbol{v} + \frac{\partial f}{\partial \boldsymbol{v}} \cdot \frac{\boldsymbol{K}}{m} = \frac{\partial f}{\partial \boldsymbol{x}} \cdot \frac{\mathrm{d}\boldsymbol{x}}{\mathrm{d}t} + \frac{\partial f}{\partial \boldsymbol{v}} \cdot \frac{\mathrm{d}\boldsymbol{v}}{\mathrm{d}t} = 0\,,$$

that is, only if f is a time-independent integral of the equations of motion of the molecule. From this, we can obtain the following most general distribution function [A-1]:

$$f = A \exp\left[-\alpha\left(\frac{2(E_{\mathrm{kin}}+E_{\mathrm{pot}})}{m} - 2\boldsymbol{v}\cdot\boldsymbol{c}\right)\right].$$

Here, (1) $\alpha = \text{constant}$ and $A = \text{constant}$,

(2) $\boldsymbol{K}$ must have a potential E_{pot}, and

(3) $\boldsymbol{c} = \boldsymbol{c}^0 + \boldsymbol{\omega}\times\boldsymbol{x}$ corresponds to the motion of a rigid body, so that the velocity vectors are everywhere tangential to the equipotential surfaces.

5. TRANSPORT PHENOMENA

The fundamental equation also provides the exact basis for treating transport phenomena.

a. Viscosity

We neglect external forces:

$$\frac{\partial f}{\partial t} + \frac{\partial f}{\partial \boldsymbol{x}}\cdot\boldsymbol{v} = \iint \mathrm{d}^3V\,\mathrm{d}^2\lambda\, wq[f(\boldsymbol{v}')f(\boldsymbol{V}') - f(\boldsymbol{v})f(\boldsymbol{V})].$$

The pressure tensor,

$$p_{ik} = \int m v_i v_k f\,\mathrm{d}^3v,$$

is to be calculated. As an example, we consider the case of a gas between two plane plates, one stationary and one moving. As a zeroth-order approximation, we can set

$$c_x(z) = C\frac{z}{d} \qquad \text{(from the macroscopic theory);}$$

then,

$$f_0(\boldsymbol{v}) = A\exp[-\alpha(\boldsymbol{v}-\boldsymbol{c})^2]$$
$$= A\exp\left[-\alpha\left[\left(v_x - C\frac{z}{d}\right)^2 + v_y^2 + v_z^2\right]\right].$$

However, this is not a solution of the fundamental equation, because of the term $(\partial f/\partial z)v_z$ on the left-hand side.

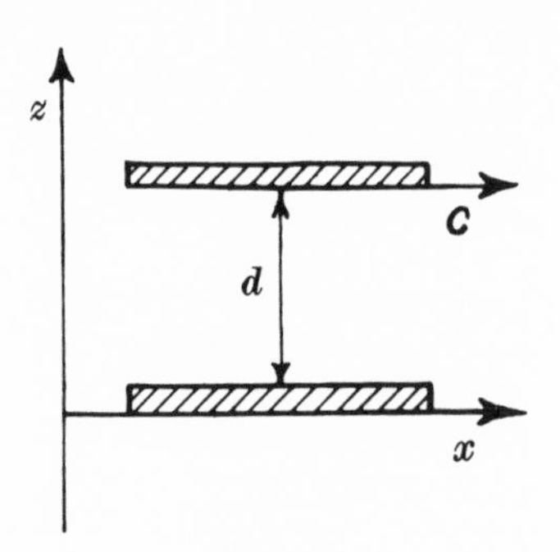

Figure 5.1

We obtain a further approximation by writing

$$f = f_0[1 + u_x u_z B(u^2) + \ldots] ,$$

with

$$u_x = v_x - C\frac{z}{d}, \qquad u_y = v_y , \qquad u_z = v_z .$$

The fundamental equation yields

$$-u_z \frac{\partial f}{\partial u_x}\frac{C}{d} = \iint \mathrm{d}^2\lambda\, \mathrm{d}^3U wq \{f_0(\boldsymbol{u}')f_0(\boldsymbol{U}')[u'_x u'_z B(u'^2) + U'_x U'_z B(U'^2)] - f_0(\boldsymbol{u})f_0(\boldsymbol{U})[u_x u_z B(u^2) + U_x U_z B(U^2)]\} .$$

It is easily verified that the dependence on the direction of the velocity which results from the spherical symmetry of the forces is identically satisfied. What remains is a complicated integral equation for $B(u^2)$. (See S. Chapman and T. G. Cowling, *Mathematical Theory of Non-Uniform Gases.*) The shearing stress can be calculated in terms of $B(u^2)$:

$$p_{xz} = m\int u_x^2 u_z^2 f_0(\boldsymbol{u}) B(u^2)\, \mathrm{d}^3u .$$

Without calculation, it is clear that the result must be

$$p_{xz} = -\eta \frac{C}{d} = -\eta \frac{\partial c_x}{\partial z} \,.$$

Remark: Our approximation makes the shearing stress linear. By using the concept of the *mean free path* l, which, of course, is implicitly contained in the integral equation, we can, on the basis of the elementary theory, suppose that the approximation will be good only if l is small compared to the dimensions of the vessel. The higher-order terms, which must be considered if this condition is not satisfied, result in deviations from the macroscopic theory.

b. *Heat conduction*

The heat conduction problem can be solved with the more general expression [A-1]

$$f(\boldsymbol{u}) = f_0(\boldsymbol{u}) \left\{ 1 + \sum_{i,k} (u_i u_k - \tfrac{1}{3} \delta_{ik} u^2) \frac{\partial c_i}{\partial x_k} B(u^2) + \sum_k u_k \frac{\partial T}{\partial x_k} A(u^2) + \dots \right\} ,$$

where $T = m/(2k\alpha)$ is the temperature and k Boltzmann's constant.

We now compare the results of the elementary theory with those of the rigorous theory of Chapman. With the elementary theory, the viscosity coefficient is given by

$$\eta = \frac{\gamma}{\pi^{\frac{3}{2}}} \frac{\sqrt{mkT}}{\sigma^2} ,$$

where $\gamma \sim 1$ is an undetermined factor; the heat conduction coefficient is given by $\varkappa = \delta \eta c_v$, where δ is another factor of order one. With the rigorous theory, we obtain

$$\eta = \frac{1}{\sigma^2} \sqrt{\frac{mkT}{\pi}} \times \frac{5}{16} \times 1.016 \,,$$

which means

$$\gamma = \frac{5.08}{16} \times \pi = 0.998 \,;$$

we also obtain

$$\varkappa = 2.522 \times \eta c_v \,,$$

which implies

$$\delta = 2.522 \,.$$

Note: (1) The form of the law of force between the molecules does not affect the lack of dependence of η on pressure; it does, however, affect the temperature dependence of η.

(2) The fact that the only tensor which enters is one whose trace is zero, i.e., $u_i u_k - \frac{1}{3}\delta_{ik} u^2$, is a result of the spherical symmetry of the forces, and it leads to the consequence that the pressure tensor is zero for spherically symmetric velocity distributions. In other words, the two constants, which are possible in principle because we calculate with tensors, reduce to one constant:

$$p_{ik} = -\eta \left\{ \left(\frac{\partial c_i}{\partial x_k} + \frac{\partial c_k}{\partial x_i} \right) - \tfrac{2}{3}\delta_{ik} \sum_l \frac{\partial c_l}{\partial x_l} \right\}, \qquad \sum_i p_{ii} = 0 \,.$$

6. MEANING OF THE *H*-THEOREM. TEMPERATURE

From Section 4 we have

$$\mathscr{H} = \iint \mathrm{d}^3x\, \mathrm{d}^3v\, (f \log f - f) = \mathscr{H}_B - N \,.$$

We now specialize to a Maxwell distribution:

$$f = A \exp[-\alpha v^2] \,, \qquad \int f\, \mathrm{d}^3 v = A \left(\frac{\pi}{\alpha}\right)^{\frac{3}{2}} = n \,,$$

so that

$$A = n \left(\frac{\alpha}{\pi}\right)^{\frac{3}{2}}, \qquad \overline{v_1^2} = \overline{v_2^2} = \overline{v_3^2} = \tfrac{1}{3}\overline{v^2} = \frac{1}{2\alpha} \,.$$

Then,

$$\begin{aligned} \mathscr{H} &= \iint \mathrm{d}^3x\, \mathrm{d}^3v\, f (\log A - \alpha v^2 - 1) = N (\log A - \alpha \overline{v^2} - 1) \\ &= N (\log n + \tfrac{3}{2} \log \alpha - \tfrac{3}{2} \log \pi - \tfrac{5}{2}) \\ &= N (\log N - \log V + \tfrac{3}{2} \log \alpha - \tfrac{3}{2} \log \pi - \tfrac{5}{2}) \,. \end{aligned}$$

Furthermore,

$$E = N \frac{m}{2} \overline{v^2} = N \frac{3m}{4\alpha},$$

$$\mathscr{H} = N \left(- \log V - \tfrac{3}{2} \log E + \left[\log N + \tfrac{3}{2} \log \left(\frac{3mN}{4\alpha} \right) - \frac{5}{2} \right] \right)$$
$$= N (- \log V - \tfrac{3}{2} \log E + \text{constant}) .$$

For the thermodynamic entropy S we have
1. S is only determined to within an additive constant,
2. $p = T(\partial S/\partial V)_E$, $1/T = (\partial S/\partial E)_V$.
For one mole of an ideal gas,

$$S = C_v \log E + R \log V + \text{constant} .$$

On the other hand, for one mole of a monatomic gas,

$$\mathscr{H} = - \frac{3L}{2} \log E - L \log V + \text{constant} ,$$

where $L =$ Loschmidt's number (Avogadro's number). Comparing, with $R/L = k$, we obtain

$$S = - k\mathscr{H} + \text{constant} ,$$

and $C_v = \frac{3}{2}R$, which confirms a previous result (see end of Sec. 1). From this it follows that

$$\left(\frac{\partial S}{\partial E} \right)_V = \frac{1}{T} = - k \left(\frac{\partial \mathscr{H}}{\partial E} \right)_V = + \frac{3Lk}{2\bar{E}} = \frac{2\alpha k}{m} ,$$

where

$$\alpha = \frac{m}{2kT}, \qquad \text{or} \qquad T = \frac{m}{2k\alpha} .$$

Remark: In thermodynamics, temperature is the primary quantity; from this, with the help of general assumptions (the first two laws), the concept of entropy is derived. On the other hand, in statistical mechanics a function $\mathscr{H}$ can be defined which has the properties of entropy; from $\mathscr{H}$ a temperature can then be defined as a secondary quantity,

$$\left(\frac{\partial \mathscr{H}}{\partial E}\right)_V = -\frac{1}{kT},$$

as was done above for a monatomic ideal gas. Using the concept of temperature, $\mathscr{H}$ can be written as

$$-\mathscr{H} = N\left(\log \frac{V}{N} + \tfrac{3}{2}\log T + \tfrac{3}{2}\log \frac{2\pi k}{m} + \frac{5}{2}\right).$$

We must not ascribe too much meaning, in the sense of Nernst's theorem, to the entropy constants which appear here, because $\mathscr{H}$ is not invariantly defined with respect to dimensions. Indeed, $f\mathrm{d}^3v\mathrm{d}^3x$ is dimensionless, but f is not; therefore, $\int\int f \log f \mathrm{d}^3v\mathrm{d}^3x$ is only defined to within an additive dimension-dependent constant, and the indeterminacy of the entropy constants remains. This question is only resolved in quantum statistics.

7. STATISTICS OF AN IDEAL GAS

We divide ($\boldsymbol{v}$, $\boldsymbol{x}$)-space into cells, each of which is of constant volume ω, and which we number with an index k;

$$N_k = \int\int_k f\,\mathrm{d}^3v\,\mathrm{d}^3x$$

is the time-varying occupation number of the kth cell. Since the N_k change discontinuously, we only consider finite time intervals τ.

We call the set of all N_k a *macrostate* of the gas. Of course, we always have

$$\sum_k N_k = N, \qquad \text{and} \qquad \sum_k N_k \varepsilon_k = E = \text{constant},$$

with $\varepsilon_k = mv_k^2/2$ for our case of a monatomic ideal gas.

A *microstate* of the gas is defined to be a set of numbers which specify in which cell each atom is located:

s = number labeling the atom,

k_s = index of the cell in which atom s is located,

(k_s) = microstate.

We have

$$\sum_{s=1}^{N} \varepsilon_{k_s} = E .$$

The macrostate is uniquely determined by the microstate, although the converse is not true. For every macrostate there are very many microstates, i.e., as many as the number of ways in which N elements can be ordered into groups of N_1, N_2, ... identical elements. This "multinomial coefficient" is well known:

$$\frac{N!}{N_1!N_2!\ldots N_k!\ldots} = \frac{N!}{\prod_k N_k!} .$$

Boltzmann's Fundamental Hypothesis: All microstates are equally probable.

Discussion: We must consider "probability" as relative (*a posteriori*) frequency of occurrence, but in what kind of ensemble? The possibilities are

1. *Time ensemble of a gas*, consisting of discrete times out of the time development of the gas,
2. *Statistical ensemble*, consisting of very *many* gases at *one* instant in time.

Boltzmann assumed that these two ensembles give the same results. The time ensemble is determined mechanically, and it can be expected that this hypothesis will not be satisfied for certain special initial states. The weight of these exceptional initial states should, however, be vanishingly small compared with that of the other initial states.

For the validity of the Boltzmann hypothesis, it is necessary that there be no integral of the equations of motion other than the energy; this is immediately clear. Sufficient conditions are less trivial.

The Boltzmann hypothesis is closely related to Gibbs's ergodic hypothesis, which apparently cannot be proved within the framework of classical mechanics.

If we accept the Boltzmann hypothesis, then it follows that the *relative* probability of the macrostate (N_k) is

$$W = \frac{N!}{\prod_k N_k!} .$$

For $N_k \gg 1$, using Stirling's formula we obtain

$$\log W = \log N! - \sum_k N_k (\log N_k - 1) .$$

a. Relation to $\mathscr{H}$

$$\begin{aligned} \mathscr{H} &= \iint f(\log f - 1)\, \mathrm{d}^3v\, \mathrm{d}^3x \\ &= \sum_k N_k \left(\log \frac{N_k}{\omega} - 1\right) \\ &= \sum_k N_k (\log N_k - 1) - N \log \omega ; \end{aligned}$$

hence,

$$\log W = \log N! - \mathscr{H} - N \log \omega .$$

The first and third terms are independent of the distribution; they compensate for the different dimension of $\mathscr{H}$.

b. Most probable distribution

Using Lagrange multipliers, we obtain the following result for the extremum:

$$\frac{\partial \log W}{\partial N_k} - \alpha\varepsilon_k + \beta = 0 , \qquad N_k = A \exp[-\alpha\varepsilon_k] .$$

That is, the Maxwell-Boltzmann energy distribution is the most frequent macrostate.

c. Theory of fluctuations in ideal gases

We consider states in the neighborhood of the Maxwell distribution:

$$N_k^0 = A \exp[-\alpha\varepsilon_k],$$

$$N_k = N_k^0 + \Delta_k, \qquad \sum_k \Delta_k = 0, \qquad \sum_k \varepsilon_k \Delta_k = 0,$$

$$\log W = \log W_0 + \sum_k \left(\frac{\partial \log W}{\partial N_k}\right)_0 \Delta_k + \tfrac{1}{2}\sum_{k,l}\left(\frac{\partial^2 \log W}{\partial N_k \partial N_l}\right)_0 \Delta_k \Delta_l + \dots .$$

The linear term vanishes because of the auxiliary conditions, and the quadratic term becomes

$$\frac{\partial^2 \log W}{\partial N_k \partial N_l} = -\frac{1}{N_k}\delta_{k,l}.$$

Thus,

$$\log W = \log W_0 - \tfrac{1}{2}\sum_k \frac{1}{N_k^0}\Delta_k^2, \qquad W = W_0 \prod_k \exp\left[-\frac{1}{2}\frac{\Delta_k^2}{N_k^0}\right].$$

We can draw the following conclusions:

1. The Maxwell distribution actually represents a maximum of W.
2. For density fluctuations, we have

$$\overline{\Delta_k^2} = N_k^0 .$$

d. Discussion of the H-theorem

After $\mathscr{H}$ has reached its minimum, it will not keep this value indefinitely; rather, fluctuations will occur which are more infrequent the larger they are. If a large fluctuation occurs at time t, it will very seldom happen that an even larger one will occur at time $t+\tau$, or that a larger one has occurred at time $t-\tau$. Because of this, time-symmetry is reestablished. What we have calculated by means of the Stosszahlansatz is not valid for a single gas; it is valid only for a statistical ensemble. With a statistical ensemble, the values of $\mathscr{H}$, necessarily discrete for a single gas, are replaced by a continuous distribution of values which satisfies the H-theorem. However, when we thermodynamically observe irreversibility, the situation is always such that the distribution of positions and velocities approaches the equilibrium distribution with overwhelming probability, although small fluctuations continually occur.

e. Generalization of the concept of entropy

The statistical view also permits us to formulate a definition of entropy for nonequilibrium states. For two states, 1 and 2, we have

$$S_2 - S_1 = k \log \frac{W_2}{W_1} ;$$

leaving the additive constant unspecified, we obtain

$$S = k \log W .$$

Because of the logarithm, and because the probabilities of independent states multiply, the additivity of entropy is maintained.

Chapter 2. General Statistical Mechanics

In the following, we no longer restrict ourselves to ideal gases. We seek a general scheme with which to derive the thermodynamic functions of state of a mechanical system from its mechanical structure.

8. PHASE SPACE AND LIOUVILLE'S THEOREM

We consider a mechanical system with N degrees of freedom which is described by its equations of motion in canonical form:

$$\dot{q}_i = \frac{\partial H}{\partial p_i},$$
$$\dot{p}_i = -\frac{\partial H}{\partial q_i}.$$

With specified initial values q_i^0 and p_i^0, the integration of these equations gives the time development of the system, geometrically represented by a curve in the $2N$-dimensional *phase space*, the *orbit in phase space*:

$$\left.\begin{aligned} q_i &= q_i(q_k^0, p_k^0, t) \\ p_i &= p_i(q_k^0, p_k^0, t) \end{aligned}\right\} \quad \text{with} \quad \left\{\begin{aligned} q_i(q_k^0, p_k^0, 0) &= q_i^0 \\ p_i(q_k^0, p_k^0, 0) &= p_i^0 . \end{aligned}\right.$$

Further, from the equations of motion it follows that

$$\dot{H} = \frac{\mathrm{d}H}{\mathrm{d}t} = \frac{\partial H}{\partial t} + \sum_k \left\{\dot{q}_k \frac{\partial H}{\partial q_k} + \dot{p}_k \frac{\partial H}{\partial p_k}\right\} = \frac{\partial H}{\partial t}.$$

In case H does not depend on time explicitly, then $\partial H/\partial t=0$, and

$$H(p, q) = E = \text{constant}\,.$$

This means that for a time-independent H the orbit in phase space lies completely on an energy hypersurface. Because the orbits in phase space are uniquely determined by the differential equations and the initial conditions, two orbits can never intersect.

Liouville's theorem

The volume of a volume element in phase space does not change in the course of time if each of its points traces out the orbit in phase space determined by the equations of motion; that is,

$$D(t, t_0) = \frac{\partial(p, q)}{\partial(p^0, q^0)} = 1\,.$$

We can prove this in two steps:

1. First we show that

$$[\partial D(t, t_0)/\partial t]_{t=t_0} = 0\,.$$

Defining $\varepsilon = t - t_0$, we can write

$$\begin{aligned} D(t, t_0) &= 1 + \varepsilon \left[\frac{\partial D}{\partial t}(t, t_0)\right]_{t=t_0} + \ldots \\ &= 1 + \varepsilon \sum_i \left(\frac{\partial \dot{p}_i}{\partial p_i^0} + \frac{\partial \dot{q}_i}{\partial q_i^0}\right)_{t=t_0} + \ldots\,. \end{aligned}$$

Hence,

$$\begin{aligned} \left[\frac{\partial D(t, t_0)}{\partial t}\right]_{t=t_0} &= \sum_i \left(\frac{\partial \dot{p}_i}{\partial p_i^0} + \frac{\partial \dot{q}_i}{\partial q_i^0}\right)_{t=t_0} \\ &= \sum_i \left(-\frac{\partial^2 H}{\partial p_i^0\, \partial q_i} + \frac{\partial^2 H}{\partial q_i^0\, \partial p_i}\right)_{t=t_0} \\ &= \sum_i \left(-\frac{\partial^2 H}{\partial p_i^0\, \partial q_i^0} + \frac{\partial^2 H}{\partial q_i^0\, \partial p_i^0}\right) = 0\,. \end{aligned}$$

2. From the multiplication law for Jacobians, we have

$$\frac{\partial D(t, t_0)}{\partial t} = \frac{\partial D(t, t_1)}{\partial t} D(t_1, t_0)\,.$$

Letting t_1 approach t, we then obtain

$$\frac{\partial D(t, t_0)}{\partial t} = \left[\frac{\partial D(t, t_1)}{\partial t}\right]_{t_1=t} D(t, t_0) .$$

Hence

$$\frac{\partial D(t, t_0)}{\partial t} = 0 , \qquad D(t, t_0) = \text{constant} .$$

Since $D(t_0, t_0) = 1$, we obtain the desired result, $D(t, t_0) = 1$.

9. MICROCANONICAL ENSEMBLE

We consider statistical ensembles with density $\varrho(p, q; t)$ in phase space; that is, in the volume element $\mathrm{d}^{2N}\Omega$ there are $\varrho(p, q; t)\,\mathrm{d}^{2N}\Omega$ ensemble points. As in hydrodynamics, let $\partial/\partial t$ represent differentiation at a fixed point, and let $\mathrm{D}/\mathrm{D}t$ represent differentiation along the mechanical orbit in phase space:

$$\frac{\mathrm{D}}{\mathrm{D}t} = \frac{\partial}{\partial t} + \sum_k \left(\dot{q}_k \frac{\partial}{\partial q_k} + \dot{p}_k \frac{\partial}{\partial p_k}\right) .$$

By definition,

$$\frac{\mathrm{D}}{\mathrm{D}t}(\varrho\, d^{2N}\Omega) = 0 .$$

Furthermore, according to Liouville's theorem,

$$\frac{\mathrm{D}}{\mathrm{D}t} d^{2N}\Omega = 0 .$$

Therefore,

$$\frac{\mathrm{D}\varrho}{\mathrm{D}t} = 0 ,$$

which may be written as

$$\frac{\partial \varrho}{\partial t} + \sum_k \left(\dot{q}_k \frac{\partial \varrho}{\partial q_k} + \dot{p}_k \frac{\partial \varrho}{\partial p_k}\right) = 0 ,$$

$$\frac{\partial \varrho}{\partial t} + \sum_k \left(\frac{\partial H}{\partial p_k}\frac{\partial \varrho}{\partial q_k} - \frac{\partial H}{\partial q_k}\frac{\partial \varrho}{\partial p_k}\right) = 0 .$$

In terms of Poisson brackets, the statement is

$$\frac{\partial \varrho}{\partial t} + [H, \varrho] = 0 \,.$$

If the distribution is stationary, we must have $\partial\varrho/\partial t = 0$, or

$$[H, \varrho] = 0 \,.$$

This is exactly the condition for a time-independent integral of the equations of motion. Thus, a stationary density distribution may only depend on time-independent integrals of the equations of motion.

The simplest case of a stationary distribution is when ϱ depends only on the energy: $\varrho = \varrho(E)$. This is trivially a stationary distribution and, as is certainly not easy to show, it is also the only such distribution which is regular (i.e., the only one whose dependence on p and q is not too discontinuous). Therefore, on purely physical grounds one can expect this case to include all reasonable densities.

Especially interesting are those densities which only depend on energy and which describe closed systems:

$$\varrho = \varrho(E) = \begin{cases} \text{constant}, & E < H < E + \mathrm{d}E \,, \\ 0, & \text{otherwise} \,. \end{cases}$$

Such a distribution is called a *microcanonical distribution.* For the systems considered here, the energy shell is generally a closed, or at least finite, hypersurface. For this reason, the volume of the energy shell is finite:

$$\int\limits_{E<H<E+\mathrm{d}E} \mathrm{d}^{2N}\Omega = \omega(E)\,\mathrm{d}E \,.$$

In this connection, there are essentially two points of view—that of Boltzmann and that of Gibbs.

a. *Boltzmann's point of view*

In principle it would suffice to consider one system and not a statistical ensemble. What is observed macroscopically are time averages of the form

$$\bar{f}^t = \lim_{T \to \infty} \frac{1}{2T} \int_{-T}^{+T} f(p, q)\, dt \, .$$

This view is based on the assumption that, apart from a vanishingly small number of exceptions, the initial conditions of a mechanical system do not influence these averages. This point of view is physically very satisfying, except that such time averages can never be computed [A-2]. For this reason the *ergodic hypothesis* is introduced. The ergodic hypothesis states that *time averages are identical with statistical averages over a microcanonical ensemble* for reasonable functions f, except for a number of initial conditions whose importance compared with that of all other initial conditions is vanishingly small.

The ergodic hypothesis can be viewed as a generalization of the fundamental hypothesis of Boltzmann (Sec. 7). Historically, the ergodic hypothesis was the assumption that the orbit in phase space comes arbitrarily close to every point in phase space during the course of time. Our assumption is stronger. As an example, consider the case in which $f = 1$ in a small region G, and $f = 0$ outside of G. Then, we find that

time average = fraction of time the system spends in G,

and

ensemble average = volume of G.

This means that the fraction of time the system spends in G is proportional to the volume of G. The original ergodic hypothesis does not go as far as this statement.

The ergodic hypothesis, which probably cannot be proved classically, can be shown to be satisfied with overwhelming probability in quantum theory.

b. Gibbs's point of view

Neither the time ensemble nor the completeness of the theory is essential. Gibbs simply assumes the canonical ensemble, which yields physically satisfactory results. In spite of the difficulties which are connected with the point of view of Boltzmann, Gibbs's concept is physically much less satisfying. For that reason, we shall adopt the point of view of Boltzmann.

The average value of an arbitrary quantity $f(p, q)$ over a microcanonical ensemble is

$$\overline{f(p,q)} = \frac{\int_\omega f \, \mathrm{d}^{2N}\Omega}{\int_\omega \mathrm{d}^{2N}\Omega} = \frac{\int_\omega f \, \mathrm{d}^{2N}\Omega}{\omega(E)\,\mathrm{d}E}\,.$$

In this formula, ω means integration over the energy shell (see p. 28)

$$E < H < E + \mathrm{d}E\,.$$

The integrals can be transformed by means of the relation

$$\int_{E<H<E+\mathrm{d}E} f \, \mathrm{d}^{2N}\Omega = \mathrm{d}E \frac{\mathrm{d}}{\mathrm{d}E} \int_{\Omega(E)} f \, \mathrm{d}^{2N}\Omega\,.$$

Here $\Omega(E)$ is the volume

$$\Omega(E) = \int_{H<E} \mathrm{d}^{2N}\Omega\,.$$

Let x_i be one of the p_k or q_k; then we have

$$\overline{x_i \frac{\partial H}{\partial x_i}} = \frac{\int_\omega x_i(\partial H/\partial x_i)\, \mathrm{d}^{2N}\Omega}{\int_\omega \mathrm{d}^{2N}\Omega} = \frac{\mathrm{d}E(\mathrm{d}/\mathrm{d}E) \int_{\Omega(E)} x_i(\partial H/\partial x_i)\,\mathrm{d}^{2N}\Omega}{\omega(E)\,\mathrm{d}E}$$

$$= \frac{(\mathrm{d}/\mathrm{d}E) \int_{\Omega(E)} x_i(\partial H/\partial x_i)\, \mathrm{d}^{2N}\Omega}{\omega(E)}\,.$$

Since $\partial E/\partial x_i = 0$, we have

$$\int_{\Omega(E)} x_i \frac{\partial H}{\partial x_i} \mathrm{d}^{2N}\Omega = -\int_{\Omega(E)} x_i \frac{\partial}{\partial x_i}(E-H)\,\mathrm{d}^{2N}\Omega .$$

Performing a partial integration, we obtain

$$-\int_{\Omega(E)} x_i \frac{\partial}{\partial x_i}(E-H)\,\mathrm{d}^{2N}\Omega = +\int_{\Omega(E)} (E-H)\,\mathrm{d}^{2N}\Omega .$$

Then,

$$\frac{\mathrm{d}}{\mathrm{d}E}\int_{\Omega(E)} x_i \frac{\partial H}{\partial x_i} \mathrm{d}^{2N}\Omega = \frac{\mathrm{d}}{\mathrm{d}E}\int_{\Omega(E)} (E-H)\,\mathrm{d}^{2N}\Omega$$

$$= \int_{\Omega(E)} \mathrm{d}^{2N}\Omega + (E-H)\Big|_{H=E} = \Omega(E) .$$

Finally,

$$\overline{x_i \frac{\partial H}{\partial x_i}} = \frac{\Omega(E)}{\omega(E)} = \frac{\Omega(E)}{\Omega'(E)} = \frac{1}{(\mathrm{d}/\mathrm{d}E)\log\Omega(E)} .$$

Therefore, this average value is independent of i. The quantity

$$\Sigma = \log\Omega = \frac{S}{k}$$

has the properties of entropy (see Part c, p. 33). In particular,

$$\left(\frac{\mathrm{d}\Sigma}{\mathrm{d}E}\right)_V \equiv \frac{1}{\Theta} = \frac{1}{kT} .$$

Therefore,

$$\overline{x_i \frac{\partial H}{\partial x_i}} = \Theta .$$

Corollary 1: Let $x_i = p_i$; then

$$\overline{p_i \frac{\partial H}{\partial p_i}} = \overline{p_i \dot{q}_i} .$$

If, as is usually the case, H is of the form

$$H = E_{\text{pot}}(q) + E_{\text{kin}}(p, q), \qquad \text{where} \qquad E_{\text{kin}} = \sum_{i,k} a_{ik}(q) p_i p_k ,$$

then

$$\sum_i p_i \frac{\partial E_{\text{kin}}}{\partial p_i} = \sum_i p_i \frac{\partial H}{\partial p_i} = 2 E_{\text{kin}} .$$

This allows us to define

$$\tfrac{1}{2} p_i \partial H / \partial p_i$$

as the kinetic energy of the ith degree of freedom. From this, the *equipartition theorem* follows:
The average kinetic energy per degree of freedom is $\frac{1}{2}kT$.

Corollary 2: Let $x_i = q_i$; then

$$\overline{q_i \frac{\partial H}{\partial q_i}} = - \overline{q_i \dot{p}_i} = - \overline{q_i K_i} = \Theta .$$

This is the *virial theorem* [A-3]:

$$- \overline{q_i K_i} = kT .$$

In addition, we can derive some support for the ergodic hypothesis. It can be trivially shown that the following time average vanishes:

$$\overline{\frac{\mathrm{d}}{\mathrm{d}t} (p_i q_i)}^{\,t} = 0 .$$

And, from the above, the statistical average over a micro-canonical ensemble is

$$\overline{\frac{\mathrm{d}}{\mathrm{d}t} (p_i q_i)} = \overline{q_i \dot{p}_i + p_i \dot{q}_i} = - \overline{q_i \frac{\partial H}{\partial q_i}} + \overline{p_i \frac{\partial H}{\partial p_i}} = 0 .$$

If this were not so, then the ergodic hypothesis would be severely shaken.

Remark 1: If, in analogy with the above, we calculate

$$\overline{x_k \frac{\partial H}{\partial x_i}},$$

for $i \neq k$, then the partial integration contributes nothing and we have the general result

$$\overline{x_k \frac{\partial H}{\partial x_i}} = \Theta \delta_{ik} .$$

Remark 2: We have used the canonical equations only to derive Liouville's theorem; since then we have not used the equations. However, Liouville's theorem is still true if we carry out a transformation whose determinant is 1. Thus, if we let

$$r_i = \sum_{k=1}^{N} \beta_{ik}(q) p_k , \qquad \text{where} \qquad |\beta_{ik}| = 1 ,$$

then all of our results are still correct if (r, q) is substituted for (p, q). In particular, we have

$$\overline{r_k \frac{\partial H}{\partial r_i}} = \Theta \delta_{ik} .$$

c. Entropy

We consider a system (E, V) which consists of two energetically independent subsystems (E_1, V_1) and (E_2, V_2). Then, the Hamiltonian separates in the same way:

$$H = H_1(x^{(1)}) + H_2(x^{(2)}) .$$

Let $W(E_1, V_1; E_2, V_2)$ be the probability that system 1 is on the energy shell of the phase space corresponding to (E_1, V_1), and that system 2 is on the energy shell of the phase space corresponding to (E_2, V_2). Then, the probability W is proportional to the product of the shell volumes

$$W(E_1, V_1; E_2, V_2)\,\mathrm{d}E_1 = \frac{\omega_1(E_1)\,\mathrm{d}E_1\,\omega(E_2)\,\mathrm{d}E_2}{\omega(E)\,\mathrm{d}E}$$

$$= \frac{\omega_1(E_1)\,\omega_2(E_2)}{\omega(E)}\,\mathrm{d}E_1\,,$$

where

$$E = E_1 + E_2$$

and the denominator assures correct normalization. For fixed E, the most probable value of E_1 is determined from $\mathrm{d}W = 0$. Also, for fixed E, $\mathrm{d}E_1 = -\mathrm{d}E_2$. Therefore,

$$\omega_2 \frac{\mathrm{d}\omega_1}{\mathrm{d}E_1} - \omega_1 \frac{\mathrm{d}\omega_2}{\mathrm{d}E_2} = 0\,, \qquad \text{or} \qquad \frac{1}{\omega_1}\frac{\mathrm{d}\omega_1}{\mathrm{d}E_1} = \frac{1}{\omega_2}\frac{\mathrm{d}\omega_2}{\mathrm{d}E_2}\,.$$

If we define

$$\sigma_{1,2} \equiv \log \omega_{1,2}\,,$$

and

$$\frac{1}{\theta_{1,2}} \equiv \frac{\mathrm{d}\sigma_{1,2}}{\mathrm{d}E_{1,2}}\,,$$

then

$$\theta_1 = \theta_2\,.$$

That is, θ has the properties of a *temperature*, σ has the properties of an *entropy*, and we are justified in writing

$$k\sigma = S + \text{constant}$$

and

$$\theta = kT\,.$$

By going to a special case, for example, that of an ideal gas, we can determine that k is the Boltzmann constant:

$$k = \frac{R}{L}\,.$$

Remark: Earlier, instead of identifying k as the Boltzmann constant, we took $k\Sigma = S$ and $\Theta = kT$. However, it can be shown that this is of no consequence, because

$$\Sigma - \sigma = O(\log N)$$

and

$$\Theta - \theta = O\left(\frac{1}{N}\right).$$

10. CANONICAL ENSEMBLE

As before, we consider a system 1+2, consisting of two noninteracting parts:

$$H = H_1(p, q) + H_2(P, Q).$$

We let the number of degrees of freedom of the "small system," 1, be much smaller than that of the "large system," 1+2. We now ask about the probability that the small system is to be found in the volume element $d^{2N_1}\Omega_1$ of its phase space, independent of the location of the second system. Letting

$$\int\limits_{E_2 < H_2 < E_2 + dE_2} d^{2N_2}\Omega = \omega_2(E_2)\, dE_2,$$

we have

$$dW = \frac{d^{2N_1}\Omega_1\, \omega_2(E_2)\, dE_2}{\int d^{2N_1}\Omega_1\, \omega_2(E_2)\, dE_2},$$

where the integration is to include the values of $E_2 = E - E_1$ for which $E_1 < E$:

$$dW = \frac{d^{2N_1}\Omega_1\, \omega_2(E - E_1)}{\int\limits_{E_1 < E} d^{2N_1}\Omega_1\, \omega_2(E - E_1)} = \frac{d^{2N_1}\Omega_1\, \omega_2(E - E_1)/\omega_2(E)}{\int\limits_{E_1 < E} d^{2N_1}\Omega_1\, \omega_2(E - E_1)/\omega_2(E)}.$$

Defining $\omega_2(E) = \exp[\sigma_2(E)]$, we can write

$$dW = \frac{d^{2N_1}\Omega_1 \exp[\sigma_2(E - E_1) - \sigma_2(E)]}{\int\limits_{E_1 < E} d^{2N_1}\Omega_1 \exp[\sigma_2(E - E_1) - \sigma_2(E)]}.$$

If we now use the assumption $N_1 \ll N_2$, then

$$\sigma_2(E - E_1) - \sigma_2(E) = -\alpha E_1[1 + O(N_1/N_2)],$$

where

$$\alpha \equiv \frac{\mathrm{d}\sigma_2(E)}{\mathrm{d}E} = \frac{1}{\theta} = \frac{1}{kT} .$$

In this way, we obtain the distribution for the small system. Writing H instead of E and dropping the index 1, we have

$$\varrho(p, q) = \frac{\exp[-\alpha H(p, q)]}{\int \exp[-\alpha H(p', q')] \mathrm{d}^{2N}\Omega'} .$$

This is the probability distribution for a *canonical ensemble.* Thus, a system in contact with a large heat reservoir corresponds to a canonical ensemble.

If we specialize to the consideration of a molecule in an ideal gas, we obtain the Maxwell-Boltzmann distribution anew.

If a canonical system can be separated into two parts, such that $H = H_1 + H_2$, then

$$\varrho = \varrho_1 \varrho_2 ,$$

which means that the subsystems are distributed canonically also. This, together with $\varrho = \varrho(H)$, is characteristic of the canonical ensemble.

11. THERMODYNAMIC FUNCTIONS OF STATE

Following Gibbs, the free energy F is defined by

$$\exp[-\alpha F] = \int \exp[-\alpha H] \, \mathrm{d}^{2N}\Omega \equiv Z ,$$

or

$$F = -kT \log Z .$$

The function Z is called the *partition function.* We can see that this is a suitable definition of the free energy by showing that the following two equations, which are required by thermodynamics, are satisfied:

$$S = -\left(\frac{\partial F}{\partial T}\right)_V, \quad E = F - T\left(\frac{\partial F}{\partial T}\right)_V .$$

Differentiating

$$\int \exp[\alpha(F-H)]\,\mathrm{d}^{2N}\Omega = 1$$

with respect to α, and using $\partial H/\partial\alpha = 0$, we obtain

$$\int\left(\alpha\frac{\partial F}{\partial\alpha} + F - H\right)\exp[\alpha(F-H)]\,\mathrm{d}^{2N}\Omega = 0\,.$$

From this, remembering the relation $\varrho = \exp[\alpha(F-H)]$, we obtain

$$\alpha\frac{\partial F}{\partial\alpha} + F - E = 0$$

or

$$-T\frac{\partial F}{\partial T} + F - E = 0\,,$$

where $\bar{H} = E$. We see that the *definition* of entropy which must be used is

$$S = -\frac{\partial F}{\partial T}\,.$$

Let us compare this entropy, whose definition is based on the canonical ensemble, with the entropy defined by means of the microcanonical ensemble:

$$\exp[-\alpha F] = \int \exp[-\alpha H]\,\mathrm{d}^{2N}\Omega\,.$$

First, we integrate over the energy shell. For this purpose, we write

$$\exp[-\alpha F] = \int \exp[-\alpha H]\,\omega(H)\,\mathrm{d}H = \int \exp[\sigma(H) - \alpha H]\,\mathrm{d}H\,.$$

The integral can be evaluated by the method of steepest descent. The integrand has its maximum at $H = E$:

$$\left(\frac{\mathrm{d}\sigma}{\mathrm{d}H}\right)_E = \alpha \qquad (E = \text{the most probable energy})\,.$$

Therefore,

$$\sigma(H)-\alpha H=\sigma(E)-\alpha E+\frac{1}{2}\left(\frac{\mathrm{d}^2\sigma}{\mathrm{d}H^2}\right)_E(H-E)^2+O\left[(H-E)^3\right],$$

which leads to

$$\exp[-\alpha F]\sim\exp[\sigma(E)-\alpha E]\int_{-\infty}^{+\infty}\exp\left[\frac{1}{2}\frac{\mathrm{d}^2\sigma}{\mathrm{d}E^2}(H-E)^2\right]\mathrm{d}H\,,$$

where the approximation is justified by the fact that the exponential falls off extremely sharply. (Of course, we must require $\mathrm{d}^2\sigma/\mathrm{d}E^2\leqslant 0$, which is in agreement with thermodynamics.) Thus,

$$\exp[-\alpha F]=\exp[\sigma(E)-\alpha E]\sqrt{\frac{2\pi}{|\mathrm{d}^2\sigma/\mathrm{d}E^2|}}$$

or

$$-\alpha F=\sigma-\alpha E+\tfrac{1}{2}\log\left(\frac{2\pi}{|\mathrm{d}^2\sigma/\mathrm{d}E^2|}\right).$$

As a result,

$$S_{\text{can}}=S_{\text{microcan}}+\frac{k}{2}\log\left(\frac{2\pi}{|\mathrm{d}^2\sigma/\mathrm{d}E^2|}\right)$$

or

$$S_{\text{can}}-S_{\text{microcan}}=k\,O(\log N)\,.$$

The canonical entropy, as we have defined it, differs from the microcanonical entropy only in an unessential way.

a. Energy distribution

By definition,

$$\int(E-H)\exp[\alpha(F-H)]\,\mathrm{d}^{2N}\Omega=0\,.$$

Differentiating with respect to α, we obtain

$$\frac{\partial E}{\partial\alpha}+\int(E-H)\left(\alpha\frac{\partial F}{\partial\alpha}+F-H\right)\exp[\alpha(F-H)]\,\mathrm{d}^{2N}\Omega=0\,,$$

or

$$\frac{\partial E}{\partial\alpha}+\int(E-H)^2\exp[\alpha(F-H)]\,\mathrm{d}^{2N}\Omega=0\,.$$

Thus, to order $O(E^2/N)$, the mean-square fluctuation of the energy about its most probable value is

$$\overline{(E-H)^2} = kT^2 \frac{\partial E}{\partial T} = kT^2 C_v ,$$

where C_v is the specific heat.

b. *Equipartition and virial theorems*

As with the microcanonical ensemble, it can be shown that

$$\overline{x_i \frac{\partial H}{\partial x_k}} = \frac{1}{\alpha} \delta_{ik} ,$$

where $x = p$ or q. This equation is an expression of the equipartition and virial theorems (the latter, of course, only when the interaction with the walls is included in H [A-3]). For example, we obtain by partial integration

$$\overline{p_i \frac{\partial H}{\partial p_i}} = \frac{\int p_i (\partial H/\partial p_i) \exp[-\alpha H]\, \mathrm{d}^{2N}\Omega}{\int \exp[-\alpha H]\, \mathrm{d}^{2N}\Omega}$$

$$= \frac{-(1/\alpha)\int p_i (\partial/\partial p_i)\,(\exp[-\alpha H])\, \mathrm{d}^{2N}\Omega}{\int \exp[-\alpha H]\, \mathrm{d}^{2N}\Omega} = \frac{1}{\alpha} .$$

c. *Processes with a variable external parameter*

Let $H = H(p, q; a)$ depend on a force parameter a. If we take the logarithm of

$$\exp[-\alpha F] = \int \exp[-\alpha H]\, \mathrm{d}^{2N}\Omega$$

and differentiate with respect to a, we obtain

$$-\alpha \frac{\partial F}{\partial a} = -\alpha \frac{\int (\partial H/\partial a)_{p,q} \exp[-\alpha H]\, \mathrm{d}^{2N}\Omega}{\int \exp[-\alpha H]\, \mathrm{d}^{2N}\Omega}$$

or

$$\left(\frac{\partial F}{\partial a}\right)_\alpha = \overline{\left(\frac{\partial H}{\partial a}\right)_{p,q}} .$$

The work done by the system is

$$A\,\delta a \equiv \delta W = -\overline{\left(\frac{\partial H}{\partial a}\right)}_{p,q}\delta a\,;$$

hence,

$$A = -\overline{\left(\frac{\partial H}{\partial a}\right)}_{p,q}.$$

For example: $a = V$, $A = p$, etc.

d. Entropy

If we write $S = k\alpha(E - F)$ and $E = \overline{H}$, then

$$\left(\frac{\partial S}{\partial a}\right)_{\alpha} = k\alpha\left[\frac{\partial E}{\partial a} - \overline{\left(\frac{\partial H}{\partial a}\right)}_{p,q}\right].$$

In addition, we can vary α:

$$\frac{1}{k}S = \alpha E + \log\int \exp[-\alpha H]\,\mathrm{d}^{2N}\Omega\,,$$

$$\frac{1}{k}\mathrm{d}S = \mathrm{d}\alpha E + \frac{-\mathrm{d}\alpha\int H\exp[-\alpha H]\,\mathrm{d}^{2N}\Omega}{\int\exp[-\alpha H]\,\mathrm{d}^{2N}\Omega} + \alpha\,\mathrm{d}E - \frac{\alpha\int(\partial H/\partial a)\exp[-\alpha H]\mathrm{d}^{2N}\Omega}{\int\exp[-\alpha H]\,\mathrm{d}^{2N}\Omega}\,\mathrm{d}a\,.$$

Since the first two terms cancel, we obtain

$$\frac{1}{k}\mathrm{d}S = \alpha\left[\mathrm{d}E - \overline{\left(\frac{\partial H}{\partial a}\right)}_{p,q}\mathrm{d}a\right]$$

or

$$\frac{1}{k}\mathrm{d}S = \alpha\left[\mathrm{d}\overline{H} - \overline{\left(\frac{\partial H}{\partial a}\right)}_{p,q}\mathrm{d}a\right].$$

The first term on the right is the total change in the internal energy, and the second is the work done by the system (the change in the energy which results from a change in a).

Because of this result, the condition for adiabatic changes, $\mathrm{d}S = 0$, shows that the internal energy changes only when work is done.

12. GENERAL DENSITY DISTRIBUTIONS

Let ϱ be a normalized density in phase space:

$$\int \varrho \, d^{2N}\Omega = 1 .$$

Then, following Gibbs, a generalized $\mathscr{H}$ function (see Sec. 6) can be defined as

$$\mathscr{H} = \int \varrho \log \varrho \, d^{2N}\Omega = \overline{\log \varrho} .$$

For the canonical distribution,

$$\varrho = \exp[\alpha(F - H)] ,$$

and

$$\mathscr{H} = \overline{\alpha(F - H)} = \alpha(F - \overline{H}) = -\frac{S}{k} ,$$

which justifies our definition. Thus, we can ascribe an entropy to every statistical ensemble, even to non-stationary ones.

a. Theorem: $\mathscr{H}$ is time-independent

From

$$\frac{\partial \varrho}{\partial t} = -\sum_k \left(\frac{\partial \varrho}{\partial q_k} \frac{\partial H}{\partial p_k} - \frac{\partial \varrho}{\partial p_k} \frac{\partial H}{\partial q_k} \right) , \quad \cdot \quad \int \frac{\partial \varrho}{\partial t} \, d^{2N}\Omega = 0 ,$$

it follows that

$$\frac{d\mathscr{H}}{dt} = \int \log \varrho \frac{\partial \varrho}{\partial t} \, d^{2N}\Omega = -\sum_k \int \log \varrho \left(\frac{\partial \varrho}{\partial q_k} \frac{\partial H}{\partial p_k} - \frac{\partial \varrho}{\partial p_k} \frac{\partial H}{\partial q_k} \right) d^{2N}\Omega$$

$$= -\sum_k \int \left\{ \frac{\partial}{\partial q_k} (\varrho \log \varrho - \varrho) \frac{\partial H}{\partial p_k} - \frac{\partial}{\partial p_k} (\varrho \log \varrho - \varrho) \frac{\partial H}{\partial q_k} \right\} d^{2N}\Omega$$

$$= -\sum_k \int \left\{ \frac{\partial}{\partial q_k} \left[(\varrho \log \varrho - \varrho) \frac{\partial H}{\partial p_k} \right] - \frac{\partial}{\partial p_k} \left[(\varrho \log \varrho - \varrho) \frac{\partial H}{\partial q_k} \right] \right\} d^{2N}\Omega .$$

This expression vanishes, as can be shown by partial integration, if the forces exerted by the wall are taken into account in H.

b. Lemma

Let

$$L(x, y) \equiv x(\log x - \log y) - x + y\,.$$

If x and y are positive, then $L(x, y) = 0$ when $x = y$, and $L(x, y) > 0$ when $x \neq y$.

Proof:

$$\frac{\partial L}{\partial x} = \log x - \log y\,,$$

$$\frac{\partial^2 L}{\partial x^2} = \frac{1}{x} \gtrless 0 \qquad \text{when } x \gtrless 0\,.$$

c. Theorem concerning the average

Let Z be a region in phase space, and let a "coarse density" be defined by

$$\bar{\varrho} = \frac{1}{\Omega}\int_Z \varrho\, \mathrm{d}^{2N}\Omega\,,$$

where $\Omega = \int_Z \mathrm{d}^{2N}\Omega$;

$$\mathscr{H} = \int_Z \varrho \log \varrho\, \mathrm{d}^{2N}\Omega\,, \qquad \overline{\mathscr{H}} = \Omega\bar{\varrho}\log\bar{\varrho}\,.$$

Since

$$\int_Z (\varrho - \bar{\varrho})\, \mathrm{d}^{2N}\Omega = 0\,,$$

therefore,

$$\mathscr{H} - \overline{\mathscr{H}} = \int_Z \varrho \log \varrho\, \mathrm{d}^{2N}\Omega - \Omega\bar{\varrho}\log\bar{\varrho} = \int_Z L(\varrho, \bar{\varrho})\, \mathrm{d}^{2N}\Omega > 0\,.$$

Thus,

$$\mathscr{H} - \overline{\mathscr{H}} > 0\,,$$

except when $\varrho = \bar{\varrho}$, in which case $\mathscr{H} = \overline{\mathscr{H}}$. It can be hoped that the $\overline{\mathscr{H}}$, which is formed from the coarse density, decreases. This can be proved only in quantum mechanics.

d. Separation theorem

Let the total system consist of two parts, 1 and 2, such that $d^{2N}\Omega = d^{2N_1}\Omega_1 d^{2N_2}\Omega_2$. Let the density contain correlations given by $\varrho(x_1, x_2)$. By integration we construct

$$\varrho_1 = \int \varrho \, d^{2N_2}\Omega_2 \quad \text{and} \quad \varrho_2 = \int \varrho \, d^{2N_1}\Omega_1 .$$

Since

$$\int \varrho_1 \varrho_2 d^{2N}\Omega = \int \varrho_1 d^{2N_1}\Omega_1 \int \varrho_2 d^{2N_2}\Omega_2 = 1 ,$$

therefore,

$$\int (\varrho - \varrho_1 \varrho_2) \, d^{2N}\Omega = 0 .$$

Let

$$\mathscr{H}_1 = \int \varrho_1 \log \varrho_1 \, d^{2N_1}\Omega_1 ,$$

$$\mathscr{H}_2 = \int \varrho_2 \log \varrho_2 \, d^{2N_2}\Omega_2 ,$$

$$\mathscr{H} = \int \varrho \log \varrho \; d^{2N}\Omega .$$

Assertion:

$$\mathscr{H} - \mathscr{H}_1 - \mathscr{H}_2 \geqslant 0 ,$$

where the equality holds only when ϱ is a product (no correlations).

Proof:

$$\mathscr{H} - \mathscr{H}_1 - \mathscr{H}_2 = \int \varrho [\log \varrho - \log \varrho_1 \varrho_2] \, d^{2N}\Omega$$

$$= \int L(\varrho, \varrho_1 \varrho_2) \, d^{2N}\Omega \geqslant 0 .$$

e. Combination theorem

We consider a collection of ensembles numbered by the index i, where

$$\int \varrho_i \, \mathrm{d}^{2N}\Omega = 1 \, .$$

Let $\varrho = \sum_i c_i \varrho_i$, where $\sum_i c_i = 1$ and $c_i > 0$. Then

$$\begin{aligned}
\Delta\mathscr{H} &= \sum_i c_i \int \varrho_i \log \varrho_i \, \mathrm{d}^{2N}\Omega - \int \varrho \log \varrho \, \mathrm{d}^{2N}\Omega \\
&= \sum_i c_i \int \varrho_i (\log \varrho_i - \log \varrho) \, \mathrm{d}^{2N}\Omega \\
&= \sum_i c_i \int L(\varrho_i, \varrho) \, \mathrm{d}^{2N}\Omega \geqslant 0 \, .
\end{aligned}$$

Thus,

$$\Delta\mathscr{H} \geqslant 0 \, ,$$

where the equality holds only if $\varrho_i = \varrho_k$ for all i and k. When the ensembles are combined, $\mathscr{H}$ decreases.

f. Minimum property

For fixed $E \equiv \int H\varrho \, \mathrm{d}^{2N}\Omega$, $\mathscr{H}$ is a minimum for the canonical ensemble.

Proof: Let $\varrho_0 = A \exp[-\alpha H]$. Then

$$\int \varrho \log \varrho_0 \, \mathrm{d}^{2N}\Omega = \log A - \alpha \int H\varrho \, \mathrm{d}^{2N}\Omega = \log A - \alpha E \, .$$

Fixing E implies the subsidiary condition

$$\int \varrho \log \varrho_0 \, \mathrm{d}^{2N}\Omega = \int \varrho_0 \log \varrho_0 \, \mathrm{d}^{2N}\Omega \, .$$

Then,

$$\Delta\mathscr{H} = \int \varrho \log \varrho \, \mathrm{d}^{2N}\Omega - \int \varrho_0 \log \varrho_0 \, \mathrm{d}^{2N}\Omega = \int L(\varrho, \varrho_0) \, \mathrm{d}^{2N}\Omega > 0 \, .$$

Q.E.D.

It should be noted that, instead of the energy, the average value of any quantity could have been specified.

13. APPLICATIONS

a. Diatomic molecules (dumbbells)

$$E_{\text{kin}} = \frac{A}{2}(\dot{\theta}^2 + \dot{\varphi}^2 \sin^2\theta),$$

$$p_\theta = \frac{\partial E_{\text{kin}}}{\partial \dot{\theta}} = A\dot{\theta},$$

$$p_\varphi = \frac{\partial E_{\text{kin}}}{\partial \dot{\varphi}} = A\,\dot{\varphi}\sin^2\theta.$$

If the potential energy is zero,

$$H = \frac{1}{2A}\left(p_\theta^2 + p_\varphi^2 \frac{1}{\sin^2\theta}\right).$$

The volume element of the phase space corresponding to θ and φ is

$$\mathrm{d}^4\Omega = \mathrm{d}p_\theta \mathrm{d}p_\varphi \mathrm{d}\theta\,\mathrm{d}\varphi\,.$$

If we define $\pi_1 = p_\theta$, $\pi_2 = p_\varphi/\sin\theta$ (angular velocities about $\boldsymbol{\vartheta}_0$ and $\boldsymbol{\varphi}_0$; see Fig. 13.1), and $u = \cos\theta$, then

$$\mathrm{d}^4\Omega = \mathrm{d}\pi_1 \mathrm{d}\pi_2 \mathrm{d}u\,\mathrm{d}\varphi$$

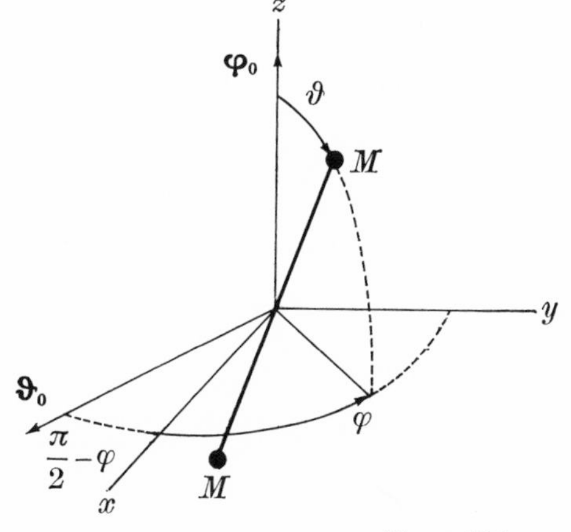

$\boldsymbol{\varphi}_0$ parallel z axis
$\boldsymbol{\vartheta}_0$ in (x, y)-plane perpendicular to molecule axis MM
φ = angle perpendicular to $\boldsymbol{\varphi}_0$
ϑ = angle perpendicular to $\boldsymbol{\vartheta}_0$

Figure 13.1

and

$$H = \frac{1}{2A}\{\pi_1^2 + \pi_2^2\} + E_{\text{pot}}(u, \varphi) .$$

Because

$$\overline{p_\theta \frac{\partial H}{\partial p_\theta}} = \overline{p_\varphi \frac{\partial H}{\partial p_\varphi}} = kT ,$$

we have

$$\overline{\frac{1}{2A}\pi_1^2} = \overline{\frac{1}{2A}\pi_2^2} = \tfrac{1}{2}kT$$

or

$$\overline{E_{\text{kin}}^{\text{rot}}} = kT .$$

This model is obeyed very well for diatomic molecules. If, further, we take into account the translational energy, $\overline{E_{\text{kin}}^{\text{tr}}} = \frac{3}{2}kT$, then

$$\overline{E_{\text{kin}}} = \tfrac{5}{2}kT .$$

It follows for the specific heats that

$$C_v = \tfrac{5}{2}R , \qquad C_p = \tfrac{7}{2}R , \qquad \text{and} \qquad \gamma = \frac{C_p}{C_v} = \frac{7}{5} = 1.4 .$$

At low temperatures the rotational degrees of freedom of the hydrogen molecule are "frozen in." This is connected with the fact that classical statistics arbitrarily counts certain degrees of freedom and does not count others. For example, here we have not counted the degree of freedom corresponding to rotation about the symmetry axis.

b. Rotating rigid bodies (general case)

We introduce the components of angular velocity along the principal axes, expressed by Euler angles:

$$\begin{aligned} u_1 &= \dot{\varphi}\sin\theta\sin\psi + \dot{\theta}\cos\psi , \\ u_2 &= \dot{\varphi}\sin\theta\cos\psi - \dot{\theta}\sin\psi , \\ u_3 &= \dot{\varphi}\cos\theta + \dot{\psi} . \end{aligned}$$

In terms of the principal moments of inertia, A_1, A_2, and A_3, we have

$$E_{\text{kin}} = \tfrac{1}{2}(A_1 u_1^2 + A_2 u_2^2 + A_3 u_3^2) .$$

Hence

$$p_\theta = A_1 \cos\psi u_1 - A_2 \sin\psi u_2 ,$$
$$p_\varphi = A_1 \sin\theta \sin\psi u_1 + A_2 \sin\theta \cos\psi u_2 + A_3 \cos\theta u_3 ,$$
$$p_\psi = A_3 u_3 ,$$

and

$$\frac{\partial(p_1, p_2, p_3)}{\partial(u_1, u_2, u_3)} = A_1 A_2 A_3 \sin\theta .$$

Then,

$$\mathrm{d}W = \text{constant} \times \exp\left[-\frac{E_{\text{kin}} + E_{\text{pot}}}{kT}\right] \mathrm{d}^6\Omega$$

$$= \text{constant} \times \exp\left[-\frac{1}{2kT}(A_1 u_1^2 + A_2 u_2^2 + A_3 u_3^2 + 2E_{\text{pot}})\right]$$
$$\times \mathrm{d}u_1 \mathrm{d}u_2 \mathrm{d}u_3 \sin\theta \, \mathrm{d}\theta \, \mathrm{d}\varphi \, \mathrm{d}\psi ,$$

and

$$\overline{\tfrac{1}{2}A_1 u_1^2} = \overline{\tfrac{1}{2}A_2 u_2^2} = \overline{\tfrac{1}{2}A_3 u_3^2} = \tfrac{1}{2}kT .$$

For polyatomic gases, we have

$$\text{rotational energy} = \tfrac{3}{2}kT ,$$

$$\text{translational energy} = \tfrac{3}{2}kT ,$$

and

$$\bar{E} = 3kT .$$

Thus,

$$C_v = 3R , \qquad C_p = 4R , \qquad \text{and} \qquad \frac{C_p}{C_v} = \frac{4}{3} = 1.33 .$$

c. Vibrating dumbbell

Introducing the reduced mass

$$\mu = Mm/(M+m) ,$$

we have

$$E = \frac{p_r^2}{2\mu} + \frac{\mu}{2}\omega_0^2(r-a)^2 + \ldots + E_{\text{rot}} .$$

For each vibrational degree of freedom, $\bar{E} = kT$. This contribution to $\bar{E}$ is in addition to the contributions from other degrees of freedom.

It is noteworthy that this energy does not in any way depend on the coupling strength (does not depend on ω_0). Therefore, for $\omega_0 \to \infty$ (rigid dumbbell), the contribution certainly does not go to zero. However, it is known experimentally that this degree of freedom generally does not enter.

d. *Solids*

We imagine that a solid consists of atoms which can oscillate about their equilibrium positions. For small oscillations the force will depend linearly on the displacement and, therefore, the potential energy will be a quadratic function of the displacement. Therefore, according to Euler's theorem,

$$\sum_k q_k \frac{\partial E_{\text{pot}}}{\partial q_k} = 2E_{\text{pot}}$$

and

$$\bar{E}_{\text{pot}} = \tfrac{1}{2}kT \cdot 3N.$$

Also,

$$\bar{E}_{\text{kin}} = \tfrac{1}{2}kT \cdot 3N,$$

which leads to

$$\bar{E} = 3RT.$$

Thus, $C_v = 3R \simeq 6$ cal/°C. This is the *law of Dulong and Petit.*

Deviations should be expected at high temperatures, since the displacements at high temperatures will not be small. However, the law is well satisfied at high temperatures, whereas, since $C_v \to 0$ as $T \to 0$, it is not satisfied at all at low temperatures. The explanation of this requires quantum statistics (see p. 82).

e. *Langevin's theory of paramagnetism*

We consider an ideal gas in a magnetic field H, each of whose molecules has a permanent magnetic moment $\boldsymbol{\mu}$.

(Exactly the same considerations are valid for dilute solutions.) We need not discuss the kinetic energy, which only depends on the π's;

$$E_{\mathrm{pot}} = -\mu H \cos\theta\,,$$

where θ is the angle between $\boldsymbol{\mu}$ and the magnetic field $\boldsymbol{H}$. Therefore,

$$W(\theta)\,\mathrm{d}\theta = \text{constant} \times \exp\left[-\frac{E_{\mathrm{pot}}}{kT}\right] \sin\theta\,\mathrm{d}\theta\,,$$

which leads to

$$\bar{M} = \overline{\mu\cos\theta} = \mu\,\frac{\int_0^{\pi} \cos\theta \exp\left[-(E_{\mathrm{pot}}/kT)\right]\sin\theta\,\mathrm{d}\theta}{\int_0^{\pi} \exp\left[-(E_{\mathrm{pot}}/kT)\right]\sin\theta\,\mathrm{d}\theta}\,.$$

From the partition function,

$$\exp\left[-\frac{F}{kT}\right] = \int_0^{\pi} \exp\left[\frac{\mu H}{kT}\cos\theta\right]\sin\theta\,\mathrm{d}\theta\,,$$

it follows that

$$\bar{M} = -\left(\frac{\partial F}{\partial H}\right)_T.$$

With $x = \mu H/kT$, we have

$$\exp\left[-\frac{F}{kT}\right] = \int_{-1}^{+1} \exp\left[\frac{\mu H}{kT}\,u\right]\mathrm{d}u$$

$$= \frac{kT}{\mu H}\left(\exp\left[\frac{\mu H}{kT}\right] - \exp\left[-\frac{\mu H}{kT}\right]\right) = 2\,\frac{\sinh x}{x}\,.$$

Therefore

$$F(H) - F(0) = -kT\log\frac{\sinh x}{x}\,,$$

$$\bar{M} = -\frac{\partial F}{\partial H} = \mu\left(\coth x - \frac{1}{x}\right).$$

Limiting cases:

(*a*) $x \gg 1$: $\qquad \bar{M} = \mu \qquad$ (saturation) .

(*b*) $x \ll 1$: $\qquad F \sim -kT\dfrac{x^2}{6} + \dots,$

$$\bar{M} = \mu\frac{x}{3} = \mu\frac{\mu H}{3kT}.$$

The susceptibility per mole is

$$\chi = \frac{1}{3}\frac{\mu^2}{kT}L = \frac{1}{3}\frac{(\mu L)^2}{RT}.$$

This is Curie's law.

Remark: It is essential to ascribe a permanent magnetic moment μ to the atom without inquiring about its origin; only quantum theory can answer the question of the origin of the permanent magnetic moment. As a matter of fact, if the magnetic moments are considered to arise from elementary amperian currents, and if the elementary particles (nuclei and electrons) are treated statistically, then the absurd result follows that paramagnetism does not exist.

For charged particles in a magnetic field $\boldsymbol{H}$, we have

$$H = \frac{1}{2m}\left(\boldsymbol{p} - \frac{e}{c}\boldsymbol{A}\right)^2,$$

where

$$\boldsymbol{H} = \operatorname{curl}\boldsymbol{A}$$

and

$$\frac{\partial \boldsymbol{A}}{\partial t} = 0.$$

The absence of paramagnetism can be verified as follows:

$$\dot{\boldsymbol{q}} = \frac{\partial H}{\partial \boldsymbol{p}} = \frac{1}{m}\left(\boldsymbol{p} - \frac{e}{c}\boldsymbol{A}\right),$$

$$\dot{p}_k = -\frac{\partial H}{\partial q_k} = \frac{1}{m}\sum_i\left(p_i - \frac{e}{c}\boldsymbol{A}_i\right)\frac{e}{c}\frac{\partial A_i}{\partial q_k},$$

$$m\ddot{q}_k = \dot{p}_k - \frac{e}{c}\sum_i\frac{\partial A_k}{\partial q_i}\dot{q}_i.$$

Hence,

$$m\ddot{q}_k = \frac{e}{c}\sum_i \dot{q}_i\left(\frac{\partial A_i}{\partial q_k} - \frac{\partial A_k}{\partial q_i}\right) = \frac{e}{c}\sum_i H_{ki}\,\dot{q}_i\ ,$$

$$m\ddot{\boldsymbol{q}} = -\frac{e}{c}\boldsymbol{H}\times\dot{\boldsymbol{q}}\ ,$$

$$m\dot{\boldsymbol{v}} = \frac{e}{c}\boldsymbol{v}\times\boldsymbol{H} \qquad (\textit{Lorentz force}).$$

If we define

$$\boldsymbol{\pi} = \boldsymbol{p} - \frac{e}{c}\boldsymbol{A} = m\dot{\boldsymbol{q}}\ ,$$

then

$$\mathrm{d}^3p\,\mathrm{d}^3q = \mathrm{d}^3\pi\,\mathrm{d}^3q\ ,$$

and

$$\begin{aligned}\mathrm{d}^3v\,W &= \text{constant}\times\exp\left[-\alpha\frac{1}{2m}\boldsymbol{\pi}^2\right]\mathrm{d}^3\pi \\ &= \text{constant}\times\exp\left[-\alpha\frac{m}{2}\boldsymbol{v}^2\right]\mathrm{d}^3v\ ,\end{aligned}$$

where W is the probability density for finding velocity $\boldsymbol{v}$. This holds whether or not we assume a potential energy between the particles. Thus, we always obtain the Maxwell distribution and, in particular, there is no magnetic moment. The reason for not having a magnetic moment is illustrated schematically in Fig. 13.2. Those particles,

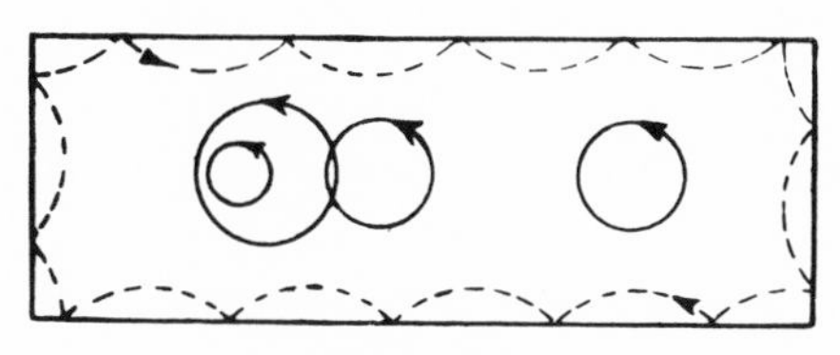

Figure 13.2

indicated by a dashed line, which are reflected by the walls, give rise to an oppositely directed current that cancels the moment produced by the other particles. *Paramagnetism is a pure quantum effect.*

f. The van der Waals gas

For the sake of simplicity, we restrict our attention to a monatomic gas. Let the potential that describes the internal force between the ith and kth molecule be $U(r_{ik})$, where $U(\infty)=0$. Then ($N=$ number of particles)

$$E_{\text{pot}} = \sum_{i<k} U(r_{ik}) \,,$$

$$H = \frac{1}{2m}\sum_{k=1}^{N} \boldsymbol{p}_k^2 + E_{\text{pot}}$$

$$\exp[-\alpha F] = \iint \exp\left[-\alpha \frac{1}{2m}\sum_{k=1}^{N}\boldsymbol{p}_k^2\right] \times \exp[-\alpha E_{\text{pot}}]\, \mathrm{d}^3p_1 \ldots \mathrm{d}^3p_N\, \mathrm{d}^3q_1 \ldots \mathrm{d}^3q_N \,.$$

If we ask just for the velocity distribution, then we again obtain the Maxwell distribution

$$\exp[-\alpha F] = (2\pi mkT)^{3N/2}\int \exp[-\alpha E_{\text{pot}}]\, \mathrm{d}^3q_1 \ldots \mathrm{d}^3q_N \,.$$

We cannot evaluate this integral exactly. However, we can make an expansion that is essentially an expansion in decreasing powers of V. Let

$$W_{ik}(r_{ik}) = \exp[-\alpha U(r_{ik})] - 1 \,; \qquad W_{ik}(\infty) = 0 \,.$$

We expand

$$\exp[-\alpha E_{\text{pot}}] = \prod_{i<k}(1+W_{ik}) = 1 + \sum_{i<k} W_{ik} + \sum_{(i<k)\neq(i'<k')} W_{ik}W_{i'k'} + \ldots \,.$$

The linear terms contribute only when two atoms are near one another; the higher-order terms contribute only when more than two atoms, or two or more pairs, "collide." For example:

$W_{12}W_{13}$: (123) together;

$W_{12}W_{34}$: (12) and (34) together at the same time.

Integrating term by term, and introducing center-of-mass coordinates, we obtain

$$\int W_{12}(r_{12})\, d^3q_1 \ldots d^3q_N = V^{N-2}\int W_{12}(r)\, d^3q_1\, d^3q_2$$

$$= V^{N-1}\int_0^\infty W(r)\cdot 4\pi r^2\, dr + \begin{matrix}\text{surface}\\ \text{terms}\end{matrix}\,.$$

Analogously,

$$\int W_{12}W_{13}\, d^{3N}q = V^{N-2}\int \ldots\,.$$

Therefore,

$$\int \exp[-\alpha E_{\text{pot}}]\, d^3q_1 \ldots d^3q_N$$

$$= V^N + V^{N-1}\binom{N}{2}\int_0^\infty 4\pi r^2 W(r)\, dr + O(V^{N-2})$$

$$\simeq V^N + V^{N-1}\frac{N^2}{2}\int_0^\infty 4\pi r^2 W(r)\, dr + O(V^{N-2})\,;$$

$$\exp[-\alpha F] = (2\pi mkT)^{3N/2}V^N$$

$$\times\left[1 + \frac{1}{V}\frac{N^2}{2}\int_0^\infty 4\pi r^2 W(r)\, dr + O\left(\frac{1}{V^2}\right)\right]\,;$$

$$-\frac{F}{kT} = \frac{3}{2}N\log(2\pi mkT) + N\log V$$

$$+ \underbrace{\frac{1}{V}\frac{N^2}{2}\int_0^\infty W(r)4\pi r^2\, dr}_{(1/V)\cdot \overline{A(T)}/kT} + O\left(\frac{1}{V^2}\right)\,.$$

This gives per mole, with $p = -\partial F/\partial V$,

$$p = \frac{RT}{V} - \frac{\overline{A(T)}}{V^2} + O\left(\frac{1}{V^3}\right)\,.$$

This is the same result obtained with the virial theorem in the kinetic theory of gases [A-3].[1]

[1] References on higher approximations: H. D. Ursell, *Proc. Cambridge. Phil. Soc.* **23**, 685 (1927); G. E. Uhlenbeck and E. Beth, *Physica* **3**, 729 (1963)

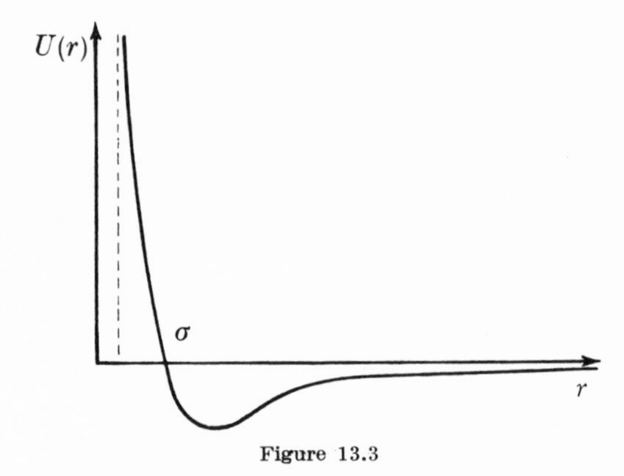

Figure 13.3

If $U(r)$ is as sketched in Fig. 13.3, then, as in kinetic theory [A-3], one finds

$$\overline{A(T)} = -RTb + a\,, \qquad b = \frac{2\pi}{3}\sigma^3 L\,,$$

$$a = -\frac{L^2}{2}\int_\sigma^\infty U(r)4\pi r^2\,\mathrm{d}r\,;$$

$$p = \frac{RT}{V}\left(1 + \frac{b}{V}\right) - \frac{a}{V^2} \qquad (\textit{van der Waals})\,.$$

14. GRAND CANONICAL ENSEMBLE

a. Homogeneous systems

In a homogeneous system of volume V and N identical molecules, we separate a subsystem of volume V_1 containing N_1 molecules by means of an imaginary wall (Fig. 14.1):

$$V = V_1 + V_2\,, \qquad N = N_1 + N_2\,.$$

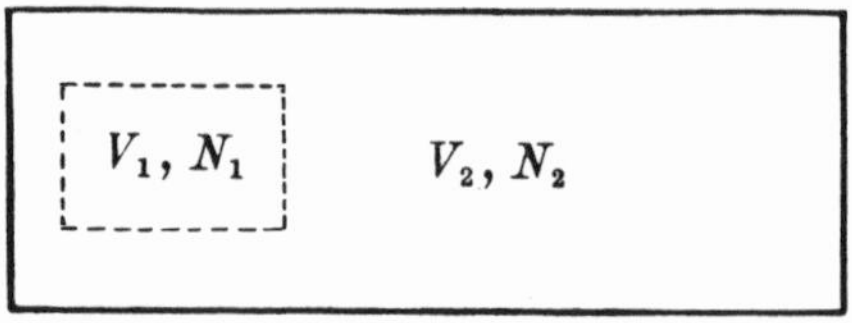

V_1, N_1 V_2, N_2

Figure 14.1

We want to determine the probability of finding N_1 molecules with coordinates $q^1, \ldots, q^{N_1}$ in V_1. First we ask for the probability of finding N_1 specific molecules in V_1. We begin with

$$W(N_1, V_1, \alpha) = \exp[+\alpha F(V, N, \alpha)] \times \int_{V_1}\int_{V_2} \exp[-\alpha H(p^1, \ldots, q^{N_1}, p^{N_1+1}, \ldots, q^{N_2})]\, d^{2N_1}\Omega_1\, d^{2N_2}\Omega_2 ,$$

where

$$\exp[-\alpha F(V, N, \alpha)] = \int_V \exp[-\alpha H(p, q)]\, d^{2N}\Omega .$$

If we assume that V_1 is large, in order that surface effects can be neglected, then we can neglect the interaction between the molecules in 1 and those in 2:

$$H = H_1(p^1, \ldots, q^{N_1}) + H_2(p^{N_1+1}, \ldots, q^{N_2}) ,$$

$$\int_{V_1}\int_{V_2} \cdots = \int_{V_1} \exp[-\alpha H_1(p^1, \ldots, q^{N_1})]\, d^{2N_1}\Omega_1 \times \int_{V_2} \exp[-\alpha H_2(p^{N_1+1}, \ldots, q^{N_2})]\, d^{2N_2}\Omega_2 .$$

Therefore, for N_1 specific molecules in V_1,

$$W(N_1, V_1, \alpha) = \exp[\alpha\{F(V, N) - F_1(V_1, N_1) - F_2(V_2, N_2)\}] ,$$

where F, F_1, and F_2 are the same function [A-4].

Now we return to the original task of determining the probability of finding any N_1 molecules in V_1:

$$W^*(N_1, V_1, \alpha) = \frac{N!}{N_1!\,N_2!} \exp[\alpha\{F(V, N) - F_1(V_1, N_1) - F_2(V_2, N_2)\}] .$$

We introduce a new function F^* by

$$\frac{1}{N!} \exp[-\alpha F] = \exp[-\alpha F^*] , \qquad F^* = F + kT \log N! .$$

Then,

$$W^*(N_1, V_1) = \exp[\alpha\{F^*(V, N) - F_1^*(V_1, N_1) - F_2^*(V_2, N - N_1\}].$$

The condition for the most probable N_1 is

$$\log W^* = \text{maximum}$$

or

$$\frac{\partial F_1^*}{\partial N_1} - \frac{\partial F_2^*}{\partial N_2} = 0 .$$

If we define $\mu = \partial F^*/\partial N$, then the condition is

$$\mu_1(V_1, N_1) = \mu_2(V_2, N_2) ,$$

where

$$N = N_1 + N_2 .$$

Fluctuations. We begin with the Taylor expansion of $\log W^*$ about the most probable value:

$$\log W^* = \log W_0 - \frac{\alpha}{2}\frac{\partial^2 F_1^*}{\partial N_1^2}(\Delta N_1)^2 - \frac{\alpha}{2}\frac{\partial^2 F_2^*}{\partial N_2^2}(\Delta N_2)^2 + \dots .$$

By using $\Delta N_1 = -\Delta N_2$, the expansion can be written as

$$\log W^* = \log W_0 - \frac{\alpha}{2}\left(\frac{\partial^2 F_1^*}{\partial N_1^2} + \frac{\partial^2 F_2^*}{\partial N_2^2}\right)(\Delta N_1)^2 + \dots .$$

The higher-order terms are small if the average values of N_1 and N_2 are large [A-5] (which was already assumed in neglecting the surface effects). Therefore, as before (compare p. 23 with the ideal gas result below, for $N_1 \ll N$),

$$\overline{(\Delta N_1)^2} = \frac{kT}{\partial^2 F_1^*/\partial N_1^2 + \partial^2 F_2^*/\partial N_2^2} .$$

Application to an ideal gas:

$$\begin{aligned} -\alpha F &= N \log V + N f(\alpha) , \\ \alpha F^* &= \log N! - N \log V - N f(\alpha) \\ &\simeq N \log N - N - N \log V - N f(\alpha) ; \end{aligned}$$

hence,

$$\alpha\mu = \log N - \log V - f(\alpha) .$$

Since $\mu_1 = \mu_2$, therefore, $N_1/V_1 = N_2/V_2$. From

$$\partial^2 F^*/\partial N^2 = 1/\alpha N$$

it then follows that

$$\overline{(\Delta N_1)^2} = \frac{1}{1/N_1 + 1/N_2} = \frac{N_1 N_2}{N} .$$

In this case, F^* is a homogeneous function of the first degree in V and N. It can be shown that this is also true in general:

$$F^*(\lambda V, \lambda N, \alpha) = \lambda F^*(V, N, \alpha) .$$

Using Euler's relation for homogeneous functions, we obtain

$$F^* = N \frac{\partial F^*}{\partial N} + V \frac{\partial F^*}{\partial V} ;$$

hence, with $\partial F^*/\partial V = -p$,

$$N\mu = F^* + pV$$

and

$$\mu = \frac{F^* + pV}{N} ,$$

which is known from thermodynamics. Also, μ is a homogeneous function of degree zero in V and N:

$$\mu(\lambda V, \lambda N, \alpha) = \mu(V, N, \alpha) .$$

Euler's relation implies

$$N \frac{\partial \mu}{\partial N} + V \frac{\partial \mu}{\partial V} = 0 ,$$

or

$$N \frac{\partial^2 F^*}{\partial N^2} + V \frac{\partial^2 F^*}{\partial V \partial N} = 0 , \qquad \frac{\partial^2 F^*}{\partial N^2} = \frac{V}{N} \frac{\partial p}{\partial N} .$$

Since p also is homogeneous of degree zero, therefore,

$$N\frac{\partial p}{\partial N}+V\frac{\partial p}{\partial V}=0\,,$$

which leads to

$$\frac{\partial^2 F^*}{\partial N^2}=-\left(\frac{V}{N}\right)^2\frac{\partial p}{\partial V}\,.$$

In order that W_0 really be a maximum, we must first have

$$\partial^2 F^*/\partial N^2>0\,,$$

which means

$$\frac{\partial p}{\partial V}<0\,.$$

This is a *stability condition.* Second, because of homogeneity, we have

$$V_1\frac{\partial p}{\partial V_1}=V_2\frac{\partial p}{\partial V_2}=V\frac{\partial p}{\partial V}\,,$$

or

$$\overline{(\Delta N_1)^2}=-\frac{kT}{(V/N)^2(\partial p/\partial V)\,V(1/V_1+1/V_2)}\,.$$

With

$$n=\frac{N}{V}\quad\text{and}\quad\frac{V}{N}\frac{\partial p}{\partial N}=\frac{1}{N}\frac{\partial p}{\partial n}\,,$$

this becomes

$$\overline{(\Delta N_1)^2}=\frac{kT}{(\partial p/\partial n)}\frac{N_1N_2}{N_1+N_2}\,.$$

Limiting case:

$$V_1\ll V_2\simeq V,\qquad N_1\ll N,$$

$$\overline{(\Delta N_1)^2}=\frac{kT}{(\partial p/\partial n)}N_1\,.$$

b. Generalization to mixtures

Let the index label the components of the mixture. Let there be $\widetilde{N}_i$ molecules of component i in the total volume $\tilde{V}$, and let there be N_i molecules of component i in the partial volume V which is separated by an imaginary wall. The associated probability is

$$W^*(N_1, N_2, \ldots, V, \alpha)$$
$$= \frac{1}{N_1!N_2!\ldots} \int_{\tilde{V}} \exp\{\alpha[F^*(\tilde{N}_1, \tilde{N}_2, \ldots, \tilde{V})$$
$$- F^*(\tilde{N}_1 - N_1, \ldots, \tilde{V} - V) - H(p^1, \ldots, q^N)]\} \mathrm{d}^{2N}\Omega,$$

where

$$F^* = F + kT \sum_k \log N_k! \, .$$

For the most probable distribution,

$$\tilde{\mu}_k \equiv \mu_k \equiv \frac{\partial F^*}{\partial N_k} \, .$$

In the following, we always assume $V \ll \tilde{V}$. Insofar as we restrict attention to the linear terms in N_1, we have

$$W^*(N_1, N_2, \ldots, V, \alpha) = \frac{1}{N_1!N_2!\ldots}$$
$$\times \int_{\tilde{V}} \exp\left[\alpha\{\Omega + \mu_1 N_1 + \mu_2 N_2 + \ldots - H(p^1, \ldots, q^N)\}\right] \mathrm{d}^{2N}\Omega \, ,$$

where we have put

$$\Omega = + \frac{\partial F^*}{\partial V} V \, .$$

For equilibrium, there is the relation

$$\frac{\partial F^*}{\partial V} = - \tilde{p} = - p \qquad \text{or} \qquad \Omega = - pV \, .$$

Consider Ω as a function of the μ_k, $\Omega(\mu_1, \mu_2, \ldots; V, T)$. For fixed T, because of the homogeneity, there is the relation

$$F^* = \sum_k \bar{N}_k \frac{\partial F^*}{\partial N_k} + V \frac{\partial F^*}{\partial V} \, .$$

Therefore,

$$\Omega = F^* - \sum_k \bar{N}_k \mu_k \, .$$

From this formula we can derive

$$-\frac{\partial \Omega}{\partial \mu_k} = \bar{N}_k \, ,$$

$$\left(\frac{\partial \Omega}{\partial V}\right)_\mu = \left(\frac{\partial F^*}{\partial V}\right)_N ,$$

$$\left(\frac{\partial \Omega}{\partial T}\right)_\mu = \left(\frac{\partial F^*}{\partial T}\right)_N .$$

The quantity Ω is a *new thermodynamic potential.*

Averages and fluctuations:

$$\sum_{N_i} W^*(N_1, N_2, \ldots, V, \alpha) = 1$$

can also be considered to be the definition of Ω:

$$\exp[-\alpha\Omega] = \sum_{N_i} \int \mathrm{d}^{2N}\Omega \, \frac{1}{\prod_i N_i!} \exp\left[\alpha\{\sum_k \mu_k N_k - H\}\right] .$$

Differentiating with respect to μ_i, we obtain

$$0 = \sum_{N_i} \left(\frac{\partial \Omega}{\partial \mu_i} + N_i\right) W^* \, .$$

Therefore, as before,

$$\bar{N}_i = -\frac{\partial \Omega}{\partial \mu_i} \, .$$

Differentiating again, we obtain

$$0 = \sum_{N_i} \left[\frac{\partial^2 \Omega}{\partial \mu_i \partial \mu_k} + \alpha \left(\frac{\partial \Omega}{\partial \mu_i} + N_i\right)\left(N_k + \frac{\partial \Omega}{\partial \mu_k}\right)\right] W^* \, .$$

From this follows

$$\overline{\Delta N^i \Delta N^k} = -kT \frac{\partial^2 \Omega}{\partial \mu_i \partial \mu_k} \, .$$

This result can also be obtained with the help of the following form of the probability, which follows for $V \ll \tilde{V}$,

$$W^* = W_0^* \exp\left[-\frac{\alpha}{2}\sum_{i,k}\frac{\partial^2 F^*}{\partial N_i \partial N_k}\Delta N_i \Delta N_k + \dots\right],$$

and with the definition

$$\overline{\Delta N_i \Delta N_k} = \int \dots \int \mathrm{d}(\Delta N)\, \Delta N_i \Delta N_k W^* .$$

The calculation of the integral is easy if the following lemma is used:

Lemma: If

$$f(x) \equiv \tfrac{1}{2}\sum_{i,k} g_{ik} x^i x^k ,$$

then

$$\overline{x^i x^k} = \frac{\int \dots \int \mathrm{d}x\, x^i x^k \exp[-\alpha f(x)]}{\int \dots \int \mathrm{d}x \exp[-\alpha f(x)]} = \frac{1}{\alpha} g^{ik},$$

where

$$\sum_\lambda g^{i\lambda} g_{\lambda k} = \delta_k^i .$$

This can be proved as follows. First we show that

$$\overline{x^i \frac{\partial f}{\partial x^k}} = \frac{1}{\alpha}\delta_k^i .$$

This follows from the relation

$$\exp[-\alpha f]\frac{\partial f}{\partial x^i} = -\frac{1}{\alpha}\frac{\partial}{\partial x^i}\exp[-\alpha f]$$

by partial integration. Therefore,

$$\overline{x^i \frac{\partial f}{\partial x^k}} = \sum_l g_{kl}\overline{x^i x^l} = \frac{1}{\alpha}\delta_k^i ,$$

which leads to

$$\overline{x^i x^l} = \frac{1}{\alpha} g^{il} . \qquad \text{Q.E.D.}$$

The above result follows with the identification

$$x_i = \Delta N_i , \qquad g_{ik} = \frac{\partial^2 F^*}{\partial N_i \partial N_k} = \frac{\partial \mu_i}{\partial N_k} , \qquad - \frac{\partial^2 \Omega}{\partial \mu_i \partial \mu_k} = \frac{\partial N_k}{\partial \mu_i} = g^{ik} .$$

By differentiating with respect to α, combined fluctuations of particles and energy can be calculated.

Chapter 3. Brownian Motion

15. INTRODUCTION

The only observable quantity is the displacement of a Brownian particle,

$$\Delta(t) = x(t) - x(0);$$

its velocity is not observable. As is always the case with irregular motions, the mean square of the displacement will be a linear function of time:

$$\overline{\Delta(t)^2} = 2Dt\,.$$

It turns out that the distribution of Δ itself is Gaussian:

$$W(\Delta)\,\mathrm{d}\Delta = \frac{1}{\sqrt{4\pi Dt}} \exp\left[-\frac{\Delta^2}{4Dt}\right] \mathrm{d}\Delta\,.$$

In the same way, rotations about an axis can be considered:

$$\overline{\Delta(\varphi)^2} = \alpha t\,.$$

It should be noted that Brownian motion cannot be used as perpetual motion of the second kind; this is because the phenomenon is exhibited by all physical devices.

In this chapter three theories of Brownian motion will be treated. They are the theories of

1. Langevin (with the virial theorem),
2. H. A. Lorentz, and
3. Einstein (comparison with diffusion).

It suffices to consider the linear case, because the motions in different directions are independent of one another.

16. LANGEVIN'S THEORY

We separate the force on a particle into an ordered part and a disordered part. The ordered part is a resistive force equal to $-W\dot{x}$. The disordered part arises from collisions between the molecules, and it will be designated by X:

$$m\ddot{x} = -W\dot{x} + X\,.$$

Because of the disorder, $\bar{X}=0$ for fixed x or $\dot{x}$. Now,

$$m\ddot{x}x = -W\dot{x}x + xX\,,$$

$$m\frac{\mathrm{d}}{\mathrm{d}t}(x\dot{x}) - m\dot{x}^2 = -W\frac{1}{2}\frac{\mathrm{d}}{\mathrm{d}t}(x^2) + xX\,.$$

If we take the statistical average of this equation over many particles, then one of the terms, $\overline{xX}$, will be zero; this is because the collisions at a fixed position occur in a completely random manner. Furthermore, the operations represented by $\mathrm{d}/\mathrm{d}t$ and the overbar (statistical average) commute:

$$m\frac{\mathrm{d}}{\mathrm{d}t}\overline{(x\dot{x})} - m\overline{\dot{x}^2} = -W\frac{\mathrm{d}}{\mathrm{d}t}\overline{(\tfrac{1}{2}x^2)}\,.$$

According to statistical mechanics,

$$m\overline{\dot{x}^2} = 2\overline{E_{\mathrm{kin}}} = kT\,.$$

Thus,

$$\frac{m}{2}\frac{\mathrm{d}^2}{\mathrm{d}t^2}\overline{(x^2)} + \tfrac{1}{2}W\frac{\mathrm{d}}{\mathrm{d}t}\overline{(x^2)} = kT\,.$$

Let us choose the origin such that $x(0)=0$. Then the solution is

$$\frac{\mathrm{d}}{\mathrm{d}t}\,(\overline{x^2}) = \frac{2kT}{W} + C\exp\left[-\frac{W}{m}\,t\right].$$

Since $W/m \cong 10^{-8}$ sec^{-1}, the stationary state is reached very quickly, and we obtain

$$\overline{x^2} = \frac{2kT}{W}\,t\,.$$

17. THEORY OF H. A. LORENTZ

Let $v \equiv \dot{x}$; then

$$m\dot{v} = -\,Wv + X\,.$$

Integrating this equation from 0 to t, we obtain

$$m(v_t - v_0) = -\,Wv_0 t + G_x\,,$$

where

$$G_x = \int_0^t X\,\mathrm{d}t'\,.$$

For sufficiently small t, using the fact that $-W\boldsymbol{v}$ is random, we obtain

$$v_t = v_0\left(1 - \frac{W}{m}\,t\right) + \frac{1}{m}\,G_x\,,$$

$$v_t^2 = v_0^2\left(1 - \frac{2W}{m}\,t\right) + 2\,\frac{G_x}{m}\,v_0\left(1 - \frac{W}{m}\,t\right) + \frac{G_x^2}{m^2} + t^2(\dots)\,.$$

If we take the statistical average of this equation and note that $\overline{v_0^2}$ must equal $\overline{v_t^2}$ and that $\overline{G_x v_0} = 0$ (as before), then we obtain

$$\overline{G_x^2} = 2Wm\overline{v_0^2}t = 2WkTt\,.$$

This is the mean square of the momentum transferred to a particle in the (short) time t.

If we define β by $\beta = 1 - Wt/m$, then

$$v_t = \beta v_0 + \frac{1}{m}\,G_x\,.$$

Consider a time interval of length nt, and let v_k and G_k be values at time kt, $k=1, 2, \ldots, n$. Then

$$\begin{aligned}
v_1 &= \beta v_0 + \frac{1}{m} G_1 , \\
v_2 &= \beta v_1 + \frac{1}{m} G_2 = \beta^2 v_0 + \frac{1}{m} (\beta G_1 + G_2) , \\
&\cdots\cdots\cdots\cdots\cdots \\
v_n &= \beta^n v_0 + \frac{1}{m} (\beta^{n-1} G_1 + \beta^{n-2} G_2 + \ldots + G_n) .
\end{aligned}$$

Letting $v_0 = G_0/m$, this can be written as

$$v_n = \frac{1}{m} \sum_{\nu=0}^{n} \beta^{n-\nu} G_\nu .$$

The quantity Δx can be written in the following way:

$$\begin{aligned}
\Delta x &= t(v_0 + v_1 + \ldots + v_{n-1}) \\
&= t \frac{G_0}{m} (1 + \beta + \ldots + \beta^{n-1}) \\
&\qquad + \frac{t}{m} G_1 (1 + \beta + \ldots + \beta^{n-2}) + \ldots + \frac{t}{m} G_{n-1} \\
&= t \frac{G_0}{m} \frac{1-\beta^n}{1-\beta} + \frac{t}{m} G_1 \frac{1-\beta^{n-1}}{1-\beta} + \ldots + \frac{t}{m} G_{n-1} \frac{1-\beta}{1-\beta} .
\end{aligned}$$

By using the relations

$$\overline{G_n v_0} = 0 \qquad \text{and} \qquad \overline{G_n G_{n'}} = \delta_{nn'} \overline{G^2} \qquad (n \neq 0) ,$$

the mean square of Δx can be written as:

$$\begin{aligned}
\overline{(\Delta x)^2} &= \frac{t^2}{m^2} \overline{G^2} \sum_{\nu=1}^{n-1} \left(\frac{1-\beta^\nu}{1-\beta} \right)^2 + \frac{t^2}{m^2} \overline{G_0^2} \left(\frac{1-\beta^n}{1-\beta} \right)^2 \\
&= \frac{t^2}{m^2} \overline{G^2} \sum_{\nu=1}^{n-1} \frac{1}{(1-\beta)^2} (1 - 2\beta^\nu + \beta^{2\nu}) + \frac{t^2}{m^2} \overline{G_0^2} \left(\frac{1-\beta^n}{1-\beta} \right)^2 \\
&= t^2 \overline{v_0^2} \frac{(1-\beta^n)^2}{(1-\beta)^2} + \frac{t^2}{m^2} \overline{G^2} \frac{1}{(1-\beta)^2} \\
&\qquad \times \left(n - 1 - 2\beta \frac{1-\beta^{n-1}}{1-\beta} + \beta^2 \frac{1-\beta^{2n-2}}{1-\beta^2} \right) .
\end{aligned}$$

We now assume $n \gg 1$. Then the term proportional to n is the most important. With $1-\beta=(W/m)t$, that term becomes

$$\overline{(\Delta x)^2} = \frac{t^2}{m^2}\, 2WkTt\, \frac{1}{(W^2/m^2)t^2}\, n = \frac{2kT}{W}\, nt\,.$$

This result is the same as that predicted by Langevin's theory.

18. EINSTEIN'S THEORY

This theory, which was put forth in 1905, is the oldest. According to statistical mechanics, the suspended particles exert a pressure

$$p = nkT\,,$$

where n is the number of particles per cm³. The diffusion coefficient D is defined phenomenologically by

$$\boldsymbol{i} = -D\,\mathrm{grad}\, n\,,$$

where $\boldsymbol{i}$ is the current density. In one dimension, this equation is

$$i = -D\frac{\mathrm{d}n}{\mathrm{d}x}\,.$$

If an external force K and a resistance $-Wv$ act on a particle, then, in the stationary case,

$$v = K/W,$$

which leads to

$$i = nv = n\frac{K}{W}\,.$$

In a stationary state in which there is current flow as a result of a pressure gradient, the force per unit volume must maintain equilibrium:

$$nK = -\frac{\mathrm{d}p}{\mathrm{d}x} = -kT\frac{\mathrm{d}n}{\mathrm{d}x}\,.$$

Therefore,

$$i = -\frac{kT}{W}\frac{\mathrm{d}n}{\mathrm{d}x} \qquad \text{and} \qquad D = \frac{kT}{W}\,.$$

Now, $\overline{(\Delta x)^2}$ will be related to D. There is a probability density $\varphi(s)$ that a particle suffers a displacement s in time τ as a result of Brownian motion. Let the number of particles between s and $s+\mathrm{d}s$ be

$$\mathrm{d}n = n\varphi(s)\,\mathrm{d}s\,.$$

Then

$$n(x,\, t+\tau)\,\mathrm{d}x = \mathrm{d}x\int_{-\infty}^{+\infty} n(x-s,\, t)\varphi(s)\,\mathrm{d}s\,.$$

For τ small,

$$n(x,t) + \frac{\partial n}{\partial t}\tau = n(x,t)\int_{-\infty}^{+\infty}\varphi(s)\,\mathrm{d}s - \frac{\partial n}{\partial x}\int_{-\infty}^{+\infty} s\varphi(s)\,\mathrm{d}s$$

$$+\frac{1}{2}\frac{\partial\, n}{\partial x^2}\int_{-\infty}^{+\infty} s^2\varphi(s)\,\mathrm{d}s + \ldots\,.$$

If we assume that the higher-order terms are small to a higher order in τ, then

$$\frac{\partial n}{\partial t} = \frac{1}{2}\frac{\overline{s^2}}{\tau}\frac{\partial\, n}{\partial x^2}\,.$$

Because of

$$\partial n/\partial t + \operatorname{div}\boldsymbol{i} = 0\,,$$

the diffusion equation becomes

$$\frac{\partial n}{\partial t} = D\frac{\partial^2 n}{\partial x^2}.$$

Comparing the two equations for $\partial n/\partial t$, we obtain

$$D = \frac{1}{2}\frac{\overline{s^2}}{\tau}, \qquad \overline{s^2} = \frac{2kT}{W}\tau .$$

Brownian rotation

This can be treated in a completely analogous way:

$$J\ddot{\varphi} = -W\dot{\varphi} + \Phi ,$$

$$\frac{J}{2}\overline{\dot{\varphi}^2} = \frac{m}{2}\overline{\dot{x}^2} = \tfrac{1}{2}kT .$$

This leads to

$$\overline{\varphi_t^2} = \frac{2kT}{W}t .$$

With

$$\Gamma_t = \int_0^t \Phi \, \mathrm{d}t ,$$

the result is, as in the theory of Lorentz,

$$\overline{\Gamma_t^2} = 2kTWt .$$

Example: Current fluctuations in a closed circuit without electromotive force.

$$L\frac{\mathrm{d}i}{\mathrm{d}t} = -Ri + X , \qquad i = \frac{\mathrm{d}q}{\mathrm{d}t} .$$

With $G_t = \int_0^t X \mathrm{d}t$, the calculation proceeds as in the theory of Lorentz:

$$\overline{G_t^2} = 2RkTt , \qquad \tfrac{1}{2}L\overline{i^2} = \tfrac{1}{2}kT ,$$

$$\overline{(\Delta q)_t^2} = \frac{2kT}{R}t .$$

Chapter 4. Quantum Statistics

19. THEORY OF BLACKBODY RADIATION

On purely thermodynamic grounds, it is possible to derive the following two laws:[1]

1. The Stefan-Boltzmann law for the total energy density:

$$u = \frac{E}{V} = aT^4 ;$$

2. Wien's displacement law for the spectral energy density:

$$\varrho_\nu = \nu^3 F\left(\frac{\nu}{T}\right) .$$

For calculating $F(\nu/T)$ we have to use statistical methods, of which there are two: the method of oscillators and the method of normal modes.

a. Method of oscillators

By an oscillator we shall mean a mechanical system (a harmonic oscillator) which can emit and absorb a specified frequency. On thermodynamic grounds, the equilibrium radiation must not depend on the details of the structure of these oscillators. Therefore, we may immedi-

[1] See W. PAULI, *Lectures in Physics: Thermodynamics and the Kinetic Theory of Gases* (M.I.T. Press, Cambridge, Mass., 1972).

ately assume that the oscillators are harmonic oscillators:

$$H = \frac{p^2}{2m} + \frac{m}{2}\,\omega^2 q^2 \,.$$

It can be shown that such an oscillator in contact with radiation has an average energy

$$\bar{E}_\nu = \frac{c^3}{8\pi\nu^2}\,\varrho_\nu \,.$$

Furthermore, according to classical statistical mechanics, the relation

$$\bar{E}_\nu = kT$$

should be valid. Substituting this into the previous formula gives

$$\varrho_\nu = \frac{8\pi\nu^2}{c^3}\,kT \,,$$

which is the Rayleigh-Jeans formula. It can be seen immediately that this formula is incorrect because $\int_0^\infty \varrho_\nu \,d\nu = \infty$ (the ultraviolet catastrophe).

In order to overcome this dilemma, Planck made the following hypotheses:

1. The energy of an oscillator can only assume discrete values:

$$E_n = E_0 + n\,\varepsilon(\nu).$$

Also, the energy of radiation is absorbed and emitted only in small quantities $n'\varepsilon$.

2. The relation

$$\bar{E} - E_0 = \frac{c^3}{8\pi\nu}\,\varrho_\nu$$

is kept. As a consequence, the statistical mechanical

partition function is now a sum instead of an integral:

$$Z = \sum_n g_n \exp[-E_n/kT] .$$

Here the third hypothesis is natural.

3. The degeneracy of state n is g_n. (That is, state n is g_n-fold degenerate.)

Corresponding to these hypotheses, the microcanonical ensemble is to be defined by

$$\text{constant} \times W_n = \begin{cases} 1, & E < E_n < E + \mathrm{d}E, \\ 0, & \text{otherwise} . \end{cases}$$

Further, if the total number of states in the energy shell (taking the weight of each state into account) is U, then the entropy is

$$S = k \log U .$$

With these new concepts, all further formulas of classical statistical mechanics are still valid. For the case of the Planck oscillator, remembering that its states cannot be degenerate,[2] we obtain

$$\begin{aligned} Z &= \exp[-E_0/kT] \sum_n \exp[-n\varepsilon/kT] \\ &= \exp[-E_0/kT] \frac{1}{1 - \exp[-\varepsilon/kT]} \end{aligned}$$

per oscillator, or

$$F = -kT \log Z = E_0 + kT \log(1 - \exp[-\varepsilon/kT]) .$$

Again taking

$$\alpha = 1/kT ,$$

we have

$$\bar{E} = \frac{\partial}{\partial\alpha}(\alpha F) = -\frac{\partial}{\partial\alpha} \log Z$$

or

$$\bar{E} = E_0 + \frac{\varepsilon}{\exp[\alpha\varepsilon] - 1} .$$

[2] See W. Pauli, *Lectures in Physics: Wave Mechanics* (M.I.T. Press, Cambridge, Mass., 1972).

Using hypothesis 2, we obtain

$$\varrho_\nu = \frac{8\pi\nu^2}{c^3} \frac{\varepsilon}{\exp[\varepsilon/kT] - 1} .$$

This has the form of the Wien law,

$$\varrho_\nu = \nu^3 F\left(\frac{\nu}{T}\right) ,$$

the validity of which is assured by its thermodynamic derivation only if

$$\varepsilon = h\nu ,$$

where h is a new universal constant, the *quantum of action.* Finally,

$$\varrho_\nu = \frac{8\pi h}{c^3} \frac{\nu^3}{\exp[h\nu/kT] - 1} ,$$

which is *Planck's law of radiation,* and

$$\bar{E} = E_0 + \frac{h\nu}{\exp[h\nu/kT] - 1} .$$

Remark: In contrast with the classical case, the dimensions of the expressions of quantum statistical mechanics are correct. Let us consider the limit $h\nu \ll kT$:

$$\exp[-h\nu/kT] = 1 - \frac{h\nu}{kT} + \frac{1}{2}\left(\frac{h\nu}{kT}\right)^2 + \cdots ,$$

so that

$$F = E_0 + kT \log\left[\frac{h\nu}{kT}\left(1 - \frac{1}{2}\frac{h\nu}{kT}\right)\right]$$

$$= E_0 + kT \log\left(\frac{h\nu}{kT}\right) - \frac{1}{2} h\nu ,$$

and

$$\bar{E} = E_0 + kT - \tfrac{1}{2} h\nu .$$

In the classical case, we found

$$\bar{E} = kT \qquad \text{and} \qquad F = kT\left(\log \frac{\nu}{kT} + \text{constant}\right) .$$

We now show that, asymptotically for large T,

$$Z_{\text{quant}} \sim Z_{\text{class}}/h \ .$$

Per oscillator,

$$Z_{\text{quant}} = \sum_n \exp[-nh\nu/kT] \exp[-E_0/kT] ,$$

$$Z_{\text{class}} = \int \mathrm{d}p \, \mathrm{d}q \exp[-H/kT] \ .$$

First we integrate over the energy shell $E<H<E+\mathrm{d}E$:

$$\omega \, \mathrm{d}E = \iint_{E<H<E+\mathrm{d}E} \mathrm{d}p \, \mathrm{d}q = \frac{\mathrm{d}\Omega}{\mathrm{d}E} \mathrm{d}E \ .$$

Using

$$H = \frac{p^2}{2m} + \frac{m}{2} \omega^2 q^2 = E ,$$

we obtain

$$\Omega = \pi \sqrt{2mE} \frac{1}{\omega} \sqrt{\frac{2}{m} E} = \frac{2\pi E}{\omega} = \frac{E}{\nu} \ .$$

Therefore,

$$\mathrm{d}\Omega = \omega \, \mathrm{d}E = \frac{\mathrm{d}E}{\nu}$$

and

$$Z_{\text{class}} = \int \exp[-E/kT] \frac{\mathrm{d}E}{\nu} ;$$

$$Z_{\text{quant}} = \sum_n \exp[-nh\nu/kT] \exp[-E_0/kT]$$

$$\cong \int \exp[-xh\nu/kT] \, \mathrm{d}x = \int \exp[-E/kT] \frac{\mathrm{d}E}{h\nu} \ .$$

Letting U be the number of states in the energy shell, we have

$$U = \omega \mathrm{d}E/h ,$$

which says that, for the linear harmonic oscillator, *the density of states in phase space is* $1/h$. This is a general result: For a system with f degrees of freedom, the density of states in phase space, asymptotically for large T, is

$$\varrho = \frac{1}{h^f} .$$

Therefore, all classical partition functions must be multiplied by h^{-f}.

For radiation the zero point energy E_0 is not important; for solids, on the other hand, it is important (see the next section). We can consider two assumptions: (1) $E_0 = 0$, or (2) $E_0 = \frac{1}{2} h\nu$. With the second assumption, the term $-\frac{1}{2} h\nu$ that appears in some of the preceding formulas is cancelled. Quantum mechanics requires assumption 2, and that assumption gives the correct result for the experimentally verified vapor pressure differences between isotopes (see the next section).

b. Method of normal modes

We consider the normal modes of a cubical cavity of length l on a side, for waves whose direction cosines are α_i. We have

$$\alpha_1^2 + \alpha_2^2 + \alpha_3^2 = 1 \qquad \text{and} \qquad \frac{\alpha_i}{\lambda} = \frac{s_i}{2l} ,$$

where s_1, s_2, and s_3 are integers greater than or equal to zero, and

$$\lambda = 2l \frac{1}{\sqrt{s_1^2 + s_2^2 + s_3^2}} .$$

For each set (λ, α_i) there are *two* normal modes in the radiation cavity that correspond to the two directions of polarization. Asymptotically, for $\lambda \ll l$, the number of normal modes in the wave number interval

$$(1/\lambda, 1/\lambda + \mathrm{d}(1/\lambda))$$

can be determined by noting that, asymptotically, the number of lattice points in a spherical shell in the first octant of s-space equals the volume of the shell:

$$N(\lambda, \lambda + \mathrm{d}\lambda) = p\,\frac{1}{8}\,4\pi R^2\,\mathrm{d}R$$
$$= p\,\frac{4\pi}{8}\,(2l)^3\,\frac{1}{\lambda^2}\,\mathrm{d}\left(\frac{1}{\lambda}\right)$$
$$= p\,\frac{4\pi V}{\lambda^2}\,\mathrm{d}\left(\frac{1}{\lambda}\right),$$

where $p =$ polarization factor, $R^2 = \sum_i s_i^2$, $R = 2l/\lambda$, and $l^3 = V$.

Note: (1) As long as $\lambda^3 \ll V$, this formula is valid for a cavity of any form. (Effects proportional to the surface area have been neglected.) (2) The formula is valid for an arbitrary dispersion law.

As a special case, consider light in vacuum:

$$\frac{1}{\lambda} = \frac{\nu}{c}, \qquad p = 2,$$

$$N(\nu, \nu + \mathrm{d}\nu) = V\,\frac{8\pi\nu^2}{c^3}\,\mathrm{d}\nu .$$

Let $\bar{E}_\nu$ be the average energy of the normal mode in thermal equilibrium. Then,

$$V\varrho_\nu \mathrm{d}\nu = N(\nu, \nu + \mathrm{d}\nu)\bar{E}_\nu, \qquad \varrho_\nu = \frac{8\pi\nu^2}{c^3}\,\bar{E}_\nu .$$

If we apply quantum theory to the normal modes, then we find

$$E_n = nh\nu,$$

which leads to

$$\bar{E}_\nu = \frac{h\nu}{\exp[h\nu/kT] - 1}.$$

From this we again obtain Planck's law,

$$\varrho_\nu = \frac{\gamma\nu^3}{\exp[h\nu/kT] - 1},$$

where

$$\gamma \equiv \frac{8\pi h}{c^3} .$$

Some implications of this law are:

1. $\dfrac{h\nu}{kT} \gg 1$: $\varrho_\nu = \gamma \nu^3 \exp\left[-\dfrac{h\nu}{kT}\right]$ (Wien's law),

2. $\dfrac{h\nu}{kT} \ll 1$: $\varrho_\nu = \dfrac{8\pi\nu^2}{c^3}\, kT$ (Rayleigh-Jeans' formula),

3. The total energy density is

$$u = \int_0^\infty \varrho_\nu \, \mathrm{d}\nu = \gamma \left(\frac{kT}{h}\right)^4 \int_0^\infty \frac{x^3\,\mathrm{d}x}{e^x - 1}$$

$$= \gamma \left(\frac{kT}{h}\right)^4 \zeta(4)\, \Gamma(4) = \gamma \left(\frac{kT}{h}\right)^4 \frac{\pi^4}{90}\, 6 .$$

Therefore,

$$u = aT^4 , \qquad a = \frac{8\pi^5 k^4}{15 c^3 h^3} \qquad \text{(Stefan-Boltzmann's law)} .$$

From this and the Planck formula, h and k can be determined separately:

$$h = 6.62 \times 10^{-27}\ \mathrm{erg \cdot sec} .$$

c. Fluctuations

Substitution of $\sum$ for $\int$ does not alter the statistical mechanical formulas:

$$\overline{(\Delta E)^2} = kT^2 \left(\frac{\partial E}{\partial T}\right)_V = -k \left(\frac{\partial E}{\partial (1/T)}\right)_V = \frac{-k}{(\partial^2 S/\partial E^2)_V} .$$

In the case of radiation, there is a special simplification. This formula is also valid for a small subvolume V, since the number of particles does not give rise to a new variable. Therefore, *the grand canonical ensemble is identical to the canonical ensemble.* Furthermore, the formula is valid for each frequency interval separately; this is because the fluctuations, like energy and entropy, are independent:

$$\overline{(\Delta E_\nu)^2} = kT^2\left(\frac{\partial E_\nu}{\partial T}\right)_V = -k\left(\frac{\partial E_\nu}{\partial(1/T)}\right)_V = \frac{-k}{(\partial^2 S_\nu/\partial E_\nu^2)_V}\,.$$

From Planck's law it follows that

$$\frac{h\nu}{k}\frac{\partial S_\nu}{\partial E_\nu} = \frac{h\nu}{kT} = \log\left(1+\frac{\gamma\nu^3}{\varrho_\nu}\right) = \log\left(\frac{\varrho_\nu}{\gamma\nu^3}+1\right) - \log\frac{\varrho_\nu}{\gamma\nu^3}\,.$$

Letting $E_\nu = \varrho_\nu \, d\nu \, V$ and $S_\nu = s_\nu \, d\nu \, V$, we obtain

$$\begin{aligned}\frac{h\nu}{k}\frac{\partial S_\nu}{\partial E_\nu} = \frac{h\nu}{k}\frac{\partial s_\nu}{\partial \varrho_\nu} &= \log\left(\frac{\varrho_\nu}{\gamma\nu^3}+1\right) - \log\frac{\varrho_\nu}{\gamma\nu^3} \\ &= \log(\varrho_\nu + \gamma\nu^3) - \log\varrho_\nu\,.\end{aligned}$$

Therefore,

$$s_\nu = \frac{k}{h\nu}\left[(\varrho_\nu+\gamma\nu^3)\log(\varrho_\nu+\gamma\nu^3) - \varrho_\nu\log\varrho_\nu - \gamma\nu^3\log\gamma\nu^3\right]$$

or

$$s_\nu = \frac{k}{h\nu}\left[(\varrho_\nu+\gamma\nu^3)\log\left(\frac{\varrho_\nu}{\gamma\nu^3}+1\right) - \varrho_\nu\log\left(\frac{\varrho_\nu}{\gamma\nu^3}\right)\right].$$

For the fluctuation we obtain

$$\overline{(\Delta E_\nu)^2} = -\left[\frac{\partial(1/kT)}{\partial E_\nu}\right]^{-1} = -\,d\nu\,V\left[\frac{\partial(1/kT)}{\partial\varrho_\nu}\right]^{-1},$$

$$\begin{aligned}\frac{\partial(1/kT)}{\partial\varrho_\nu} &= \frac{1}{h\nu}\frac{\partial}{\partial\varrho_\nu}\left[\log(\varrho_\nu+\gamma\nu^3) - \log\varrho_\nu\right] \\ &= \frac{1}{h\nu}\left[\frac{1}{\varrho_\nu+\gamma\nu^3} - \frac{1}{\varrho_\nu}\right] = \frac{1}{h\nu}\frac{-\gamma\nu^3}{\varrho_\nu(\varrho_\nu+\gamma\nu^3)}\,.\end{aligned}$$

Therefore,

$$\overline{(\Delta E_\nu)^2} = \left(\varrho_\nu h\nu + \varrho_\nu^2\frac{h\nu}{\gamma\nu^3}\right)d\nu\,V$$

or

$$\overline{(\Delta E_\nu)^2} = \left[\varrho_\nu h\nu + \frac{c^3}{8\pi\nu^2}\varrho_\nu^2\right]d\nu\,V.$$

This formula is due to Einstein. For $h=0$ (Rayleigh-Jeans), only the second term occurs. It represents interference fluctuations.[3] Only the first term occurs if we start from Wien's law, or if we consider light as a classical ideal photon gas; in the latter case, we take a photon to be a corpuscle of energy $h\nu$ which moves with the velocity of light.

In that case, let U equal the number of photons between ν and $\nu+d\nu$; $E_\nu = Uh\nu$. Then, according to p. 23,

$$\overline{(\Delta E_\nu)^2} = (h\nu)^2 \overline{\Delta U^2} = (h\nu)^2 U = h\nu E_\nu \, ,$$

which is the first term of the previous formula.

Another way of writing Einstein's formula is the following:

The number of normal modes is $\qquad N = \dfrac{8\pi\nu^2 \, d\nu}{c^3} V$.

The energy per normal mode is e_ν: $\qquad E_\nu = e_\nu N$.

Thus

$$\overline{(\Delta E)^2} = N e_\nu (h\nu + e_\nu) \, .$$

d. Radiation in a dispersive medium:

Let $n = n(\lambda)$ be the refractive index, $v = c/n$ the phase velocity,

$$U = \frac{d\nu}{d(1/\lambda)} = \frac{c}{n}\left(1 + \frac{\lambda}{n}\frac{dn}{d\lambda}\right)$$

the group velocity, and $\lambda = c/n\nu$ the wavelength. Then, the number of normal modes between ν and $\nu+d\nu$ is

$$\begin{aligned} N &= V \frac{8\pi}{\lambda^2} \, d\left(\frac{1}{\lambda}\right) \\ &= V \frac{8\pi\nu^2 \, d\nu}{c^3} \frac{n^3}{1 + (\lambda/n)(dn/d\lambda)} \\ &= V \frac{8\pi\nu^2 \, d\nu}{v^2 U} \, . \end{aligned}$$

[3] See H. A. Lorentz, *Les Théories Statistiques en Thermodynamique* (B. G. Teubner, Leipzig, 1916).

Therefore,

$$\varrho_\nu = (\varrho_\nu)_{\text{vacuum}} \frac{n^3}{1 + (\lambda/n)(\mathrm{d}n/\mathrm{d}\lambda)} = (\varrho_\nu)_{\text{vacuum}} \frac{c^3}{v^2 U} .$$

20. THEORY OF SOLIDS

The model of a solid that we use is that of N coupled mass points (atoms). Such a lattice has $3N$ normal modes. It can be shown that these vibrations always have the form of waves because of the existence of a translation group which is required by the lattice structure. For example, for a cubic lattice whose lattice constant is d, we have

$$\boldsymbol{u}_{n_1 n_2 n_3} = \boldsymbol{A} \exp \left[2\pi i \left\{ \frac{d}{\lambda} (n_1 \alpha_1 + n_2 \alpha_2 + n_3 \alpha_3 - \nu t) \right\} \right] .$$

Conditions on the α_i, analogous to those in Section 19, are

$$\frac{2\pi}{\lambda} \alpha_i l = s_i .$$

A characteristic difference in this case is that there are only a finite number of normal modes. However, for $l \gg \lambda \gg d$, the system is practically a continuum, and its normal modes are the elastic vibrations. That is,

$$\nu = \frac{1}{\lambda} v_{t,l} ,$$

where v_t is the velocity of the transverse waves ($p = 2$), and v_l is the velocity of the longitudinal waves ($p = 1$). If

$$\frac{3}{v^3} \equiv \frac{2}{v_t^3} + \frac{1}{v_l^3} ,$$

then the number of normal modes in the frequency interval $\mathrm{d}\nu$ is

$$\mathcal{N} = V \frac{12\pi}{v^3} \nu^2 \mathrm{d}\nu .$$

For small λ these considerations are invalid. In that case, there is a dispersion, which depends on the specific lattice structure, and which is such that the total number of normal modes is $3N$. Following Debye, it is a sufficient approximation to keep our distribution up to some ν_D and to cut it off at that frequency:

$$\frac{12\pi V}{v^3}\int_0^{\nu_D} \nu^2 \, d\nu = 3N$$

implies

$$\nu_D^3 = \frac{3Nv^3}{4\pi V}.$$

Eliminating v, we obtain

$$\mathcal{N} = \frac{9N}{\nu_D^3}\,\nu^2 \, d\nu.$$

Using the quantum theory of the normal modes, together with statistical mechanics, we obtain

$$\bar{E}_\nu = E_{0\nu} + \frac{h\nu}{\exp[h\nu/kT]-1}.$$

Therefore,

$$\sum_\nu \bar{E}_\nu \mathcal{N} = \bar{E} = \frac{9N}{\nu_D^3}\int_0^{\nu_D} \nu^2 \, d\nu \left(E_{0\nu} + \frac{h\nu}{\exp[h\nu/kT]-1}\right).$$

a. High temperatures: $h\nu_D/kT \ll 1$

Let us define $\widetilde{f(\nu)}$, for an arbitrary function $f(\nu)$, by

$$\widetilde{f(\nu)} \equiv \frac{3}{\nu_D^3}\int_0^{\nu_D} \nu^2 f(\nu) \, d\nu .$$

Then,

$$\bar{E} = 3N\left(\widetilde{E}_{0\nu} + \overline{\frac{h\nu}{\exp[h\nu/kT]-1}}\right).$$

For $h\nu_D/kT \ll 1$, we have

$$\bar{E} = 3NkT + 3N\left(\widetilde{E}_{0\nu} - \frac{\widetilde{h\nu}}{2}\right).$$

The free energy is

$$F^* = \sum_\nu F_\nu \mathcal{N} = \frac{9N}{\nu_D^3} \int_0^{\nu_D} \nu^2 \, d\nu \left\{ E_{0\nu} + kT \log \left(1 - \exp\left[-\frac{h\nu}{kT}\right]\right)\right\}.$$

For $h\nu_D/kT \ll 1$, we have

$$F^* = 3NkT \log\left(\frac{h\nu}{kT}\right) + 3N\left(\widetilde{E}_{0\nu} - \frac{h\widetilde{\nu}}{2}\right).$$

b. Low temperatures: $h\nu_D/kT \gg 1$

Because of the very rapid exponential decay, only the smallest values of ν contribute. Therefore, we may integrate from 0 to ∞:

$$\bar{E} - E_0 = 3N \frac{3}{\nu_D^3} \int_0^\infty \frac{h\nu^3 \, d\nu}{\exp[h\nu/kT] - 1} = \frac{9N}{\nu_D^3}\left(\frac{kT}{h}\right)^4 h \frac{\pi^4}{15},$$

where $E_0 = 3N\widetilde{E}_{0\nu}$.

With $\Theta_D \equiv \dfrac{h\nu_D}{k}$ and $Nk = R$, the result is

$$\bar{E} - E_0 = \frac{3\pi^4}{5} \frac{RT^4}{\Theta_D^3}.$$

Therefore,

$$\frac{C_v}{R} = \frac{12\pi^4}{5}\left(\frac{T}{\Theta_D}\right)^3 \qquad (T^3 \text{ law}).$$

c. General case

Define

$$F(x) = \frac{3}{x^3} \int_0^x \frac{\xi^3 \, d\xi}{e^\xi - 1}.$$

For the two extreme values of x, we have

$$x \ll 1: \qquad F(x) \simeq 1,$$

$$x \gg 1: \qquad F(x) \simeq (3/x^3)(\pi^4/15).$$

In terms of F, we can write $\bar{E}-E_0$ as

$$\bar{E}-E_0 = 3RTF\left(\frac{\Theta_D}{T}\right).$$

Neither limiting case depends on our arbitrary definition of ν_D. At low temperatures, the large values of ν are meaningless, and, at high temperatures, we simply have equipartition. It is for this reason that the Debye approximation is a good one.

d. Free energy (neglecting the zero-point energy)

$$F^* = 3NkT \log(1-\exp[-h\nu/kT])$$

$$= 3NkT\frac{3}{\nu_D^3}\int_0^{\nu_D} \log(1-\exp[-h\nu/kT])\,\nu^2\,d\nu .$$

Defining

$$\frac{h\nu}{kT} \equiv y \quad \text{and} \quad \frac{h\nu_D}{kT} \equiv x = \frac{\Theta_D}{T},$$

and integrating by parts, we obtain

$$F^* = 3NkT\frac{3}{x^3}\int_0^x \log(1-\exp[-y])\,y^2\,dy = 3NkT\,G(x) .$$

The function $G(x)$ is defined as follows:

$$G(x) = \frac{3}{x^3}\int_0^x \log(1-\exp[-y])\,y^2\,dy$$

$$= \frac{3}{x^3}\log(1-\exp[-y])\frac{y^3}{3}\Bigg|_0^x - \frac{3}{x^3}\int_0^x \frac{y^3}{3}\frac{\exp[-y]}{1-\exp[-y]}\,dy$$

$$= \log(1-\exp[-x]) - \frac{1}{3}\frac{3}{x^3}\int_0^x \frac{y^3}{e^y-1}\,dy$$

$$= \log(1-\exp[-x]) - \tfrac{1}{3}F(x) ,$$

$$xG'(x) = F(x) .$$

It follows that

$$F^* = 3NkT\,G\left(\frac{\Theta_D}{T}\right) = 3NkT\left[\log(1-\exp[-\Theta_D/T]) - \frac{1}{3}F\left(\frac{\Theta_D}{T}\right)\right].$$

For $T \ll \Theta_D$, which means $x \gg 1$,

$$G(x) \sim -\tfrac{1}{3}F(x) \sim -\frac{\pi^4}{15}\frac{1}{x^3}.$$

Therefore,

$$F^* = -\frac{\pi^4}{5}Nk\frac{T^4}{\Theta_D^3} + E_0 .$$

e. Entropy

In the approximation $T \ll \Theta_D$,

$$S^* = \frac{4\pi^4}{5}R\left(\frac{T}{\Theta_D}\right)^3.$$

Remarks: (1) The meaning of the asterisk in the symbol F^* is that F^* is the free energy defined by $F + kT\log N!$ (see p. 55). That is, it does not contain the term which arises from interchanging the molecules. For this reason, S^* is zero at zero temperature, corresponding to Nernst's theorem (see p. 93).
(2) At low temperatures the solid is analogous to the blackbody, as can be seen by replacing the Stefan-Boltzmann constant a by

$$3\pi^4R/(5V\Theta_D^3).$$

f. Zero-point energy, E_0

If we divide the volumes of the elementary cells in phase space by h^3, then we obtain the following results for an ideal, monatomic gas:[4]

$$\mu_g = \frac{\partial F^*}{\partial N_g} = kT(\log p_g - \tfrac{5}{2}\log T - i),$$

$$p_g = \frac{kTN_g}{V_g},$$

$$i = \log[(2\pi m)^{\frac{3}{2}}k^{\frac{5}{2}}h^{-3}].$$

[4] See W. Pauli, *Thermodynamics.*

Let $e_0 = E_0/N_{\text{solid}}$ = zero-point energy per oscillator for a solid. Then, for $T \gg \Theta_D$,

$$\mu_{\text{solid}} = 3kT \log \widetilde{\frac{h\nu}{kT}} - \tfrac{3}{2} h\tilde{\nu} + e_0 \ .$$

The energy e_0 consists of two parts: (1) the heat of vaporization, $-\lambda_0$, and (2) the zero-point energy of an oscillator, $\tilde{e}_{0\nu}$.

$$e_0 = -\lambda_0 + \tilde{e}_{0\nu} \ .$$

The vapor pressure follows from $\mu_{\text{solid}} = \mu_g$:

$$\log p = + \tfrac{5}{2} \log T + i + 3 \log \widetilde{\frac{h\nu}{k}} - \frac{\lambda_0}{kT} + \frac{1}{kT} (\tilde{e}_{0\nu} - \tfrac{3}{2} h\tilde{\nu}) \ .$$

For different isotopes, the values of ν are different but the values of λ_0 are the same:

$$\lambda_0' = \lambda_0 \ , \quad \nu_D'/\nu_D = \sqrt{M/M'} \quad \left(\text{since } E_{\text{pot}} = \tfrac{1}{2} m (2\pi\nu)^2 r^2 = E_{\text{pot}}'\right) .$$

Different isotopes would have different vapor pressures at high temperatures if $\tilde{e}_{0\nu} \neq \frac{3}{2} h\tilde{\nu}$. Because this is experimentally not the case, one concludes that [A-6]

$$e_{0\nu} = \tfrac{3}{2} h\nu$$

for a three-dimensional harmonic oscillator. Again, this is in agreement with quantum mechanics. Hence, for $\Theta_D \ll T$,

$$\log p = + \tfrac{5}{2} \log T + i + 3 \log \widetilde{\frac{h\nu}{k}} - \frac{\lambda_0}{kT} \ .$$

For $\Theta_D \gg T$, because $\Theta_D \sqrt{M} = \Theta_D' \sqrt{M'}$, this result is incorrect.

21. ADIABATIC INVARIANTS

For adiabatic compression of a radiation cavity, it can be shown that [A-7]

$$\frac{E_\nu}{\nu} = \frac{E_{\nu'}'}{\nu'} \ .$$

Ehrenfest made the general postulate that the quantum state of a system is not altered as a result of adiabatic changes. In agreement with this postulate, the quantum conditions would be stated as

$$nh = f(E, a_1, a_2, \ldots) ,$$

where f is an adiabatic invariant.

a. Properties of adiabatic invariants

Let the Hamiltonian function be $H(p, q; a_1, a_2, \ldots)$, where the $\boldsymbol{a}_i$ are adiabatically variable parameters:

$$\dot{p} = -\left(\frac{\partial H}{\partial q}\right)_a ,$$

$$\dot{q} = +\left(\frac{\partial H}{\partial p}\right)_a .$$

(The $\dot{a}_i$ do not enter here because they are supposed to be small.) Therefore,

$$\frac{\mathrm{d}H}{\mathrm{d}t} = \frac{\mathrm{d}E}{\mathrm{d}t} = \sum_i \left(\frac{\partial H}{\partial a_i}\right)_{p,q} \dot{a}_i .$$

For periodic motions

$$\overline{\frac{\mathrm{d}E}{\mathrm{d}t}} = \sum_i \overline{\left(\frac{\partial H}{\partial a_i}\right)}_{p,q} \dot{a}_i .$$

Since f is an adiabatic invariant,

$$\overline{\mathrm{d}f/\mathrm{d}t} = 0 ,$$

and

$$\frac{\mathrm{d}f}{\mathrm{d}t} = \frac{\partial f}{\partial E}\frac{\mathrm{d}E}{\mathrm{d}t} + \sum_i \frac{\partial f}{\partial a_i} \dot{a}_i ;$$

therefore, the conditions for adiabatic invariance are

$$\left(\frac{\partial f}{\partial E}\right)\overline{\left(\frac{\partial H}{\partial a_i}\right)}_{p,q} + \left(\frac{\partial f}{\partial a_i}\right) = 0 .$$

b. Examples:

1. *Linear oscillator*

$$\left.\begin{aligned} H &= \frac{p^2}{2m} + \frac{m}{2}\omega^2 q^2 \\ a_1 &= \omega\,, \quad a_2 = m \end{aligned}\right\} \qquad \frac{\partial H}{\partial \omega} = m\omega q^2 = \frac{2E_{\text{pot}}}{\omega}\,,$$

$$2\,\bar{E}_{\text{kin}} = \overline{q\,\frac{\partial H}{\partial q}} = \overline{p\,\frac{\partial H}{\partial p}} = 2\,\bar{E}_{\text{pot}}\,, \qquad \bar{E}_{\text{kin}} = \bar{E}_{\text{pot}} = \tfrac{1}{2}E\,.$$

Therefore,

$$\overline{\frac{\partial H}{\partial \omega}} = \frac{2\bar{E}_{\text{pot}}}{\omega} = \frac{E}{\omega}\,.$$

It is easy to verify that $f = E/\omega$ is an adiabatic invariant:

$$\frac{\partial H}{\partial m} = -\frac{p^2}{2m^2} + \frac{1}{2}\omega^2 q^2 = \frac{1}{m}(-E_{\text{kin}} + E_{\text{pot}}) = 0\,.$$

2. *Nonrelativistic particle on a line*

$$H = p^2/2m\,.$$

The force on the wall (Fig. 21.1) is

$$K = \frac{v}{2l}\,2mv = \frac{mv^2}{l} = \frac{2E}{l}\,,$$

$$\delta E = -K\,\delta l = -\frac{2E\,\delta l}{l}\,.$$

Therefore,

$$El^2 = \text{constant}\,.$$

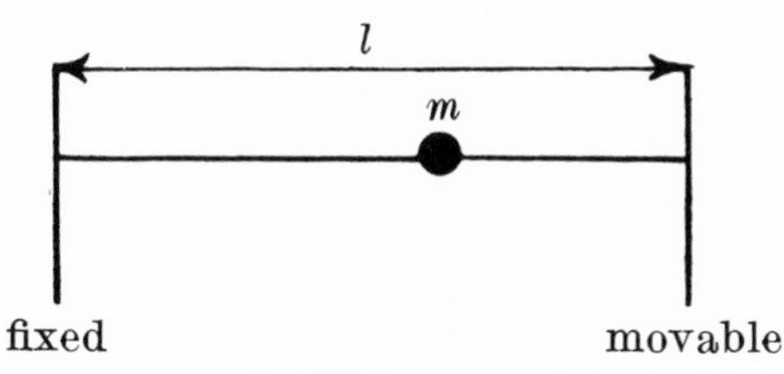

Figure 21.1

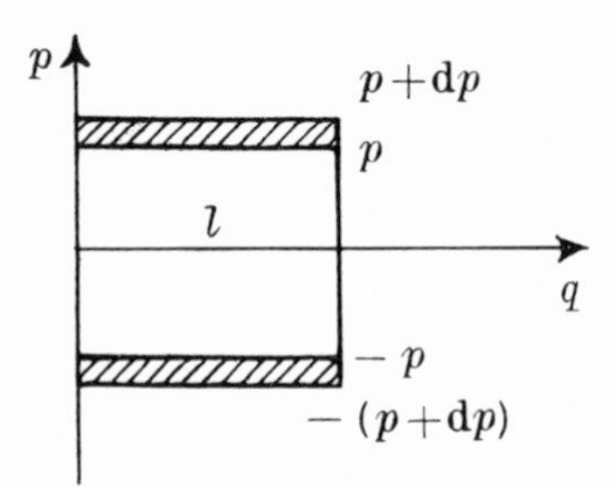

Figure 21.2

Varying the mass, we obtain

$$\delta E = -E\,\delta m/m$$

or

$$Em = \text{constant}\ .$$

Therefore,

$$Eml^2 = \text{constant}\ .$$

The energy shell of the system is given by (see Fig. 21.2)

$$\mathrm{d}\Omega = 2l\,\mathrm{d}p\ .$$

However, for $nh = 2l|p|$, the density of states in phase space is $1/h$. Therefore,

$$E_n = \frac{p^2}{2m} = \frac{n^2h^2}{8ml^2}\ .$$

From the previous considerations, this is indeed adiabatically invariant. The most general way of satisfying both conditions is obtained by replacing n by $f(n)$, where

$$\lim_{u\to\infty}\frac{f(n)}{n} = 1\ .$$

3. *Relativistic particle on a line*

$$K = 2pv/2l\ ,$$

$$v\,\delta p = \delta E = -K\,\delta l = -\frac{v}{2l}2p\,\delta l\ , \qquad \frac{\delta p}{p} + \frac{\delta l}{l} = 0\ .$$

Therefore,

$$pl = \text{constant} .$$

The conditions for adiabatic invariance and the condition on the density of states are still satisfied, because $2l|p| = nh$. Only the energy eigenvalues change:

$$E_n = c\sqrt{m^2c^2 + \frac{n^2h^2}{4l^2}} .$$

In the limiting case $m \to 0$, we have $v = c$ and $E/p = c$. Therefore,

$$E_n = c\,\frac{nh}{2l}$$

(as for the radiation cavity), and

$$El = \text{constant} .$$

For radiation,

$$\lambda = 2l\,\frac{1}{n} \quad \text{or} \quad \frac{1}{\lambda} = \frac{n}{2l} .$$

By comparison with the limiting relativistic case,

$$p = \frac{h}{\lambda} \quad \text{and} \quad E = \frac{hc}{\lambda} = h\nu .$$

These equations were the point of departure for de Broglie.

22. VAPOR PRESSURE. NERNST'S THEOREM

For a *monatomic* gas, we have (see Section 20)

$$\mu = kT\{\log p - \tfrac{5}{2}\log T - i\} + E_0 ,$$

$$i = \log\left(\frac{(2\pi m)^{\frac{3}{2}} k^{\frac{5}{2}}}{h^3}\right) ,$$

where E_0 is the zero-point energy per atom.

For a *diatomic* gas, we can no longer calculate classically, because the rotational degree of freedom about the symmetry axis must usually be considered to be frozen in. On the other hand, at room temperature we may usually calculate with the other two degrees of freedom classically. In that case, we have per molecule:

$$Z_{\text{rot}} = \frac{1}{h^2}\int \exp\left[-\frac{1}{kT}\frac{1}{2A}(\pi_1^2+\pi_2^2)\right] d\pi_1 d\pi_2 \sin\theta\, d\theta\, d\varphi$$

$$= \frac{8\pi^2 AkT}{h^2}\,.$$

For *tri-* and *poly-atomic* gases, we obtain

$$Z_{\text{rot}} = \frac{1}{h^3}\int \exp\left[-\frac{1}{2kT}\left(\frac{\pi_1^2}{A_1}+\frac{\pi_2^2}{A_2}+\frac{\pi_3^2}{A_3}\right)\right] d\pi_1 d\pi_2 d\pi_3 \sin\theta\, d\theta\, d\varphi\, d\psi$$

$$= \frac{1}{h^3}(2\pi kT)^{\frac{3}{2}}(A_1A_2A_3)^{\frac{1}{2}} \times 4\pi \times 2\pi\,.$$

For molecules with two or more identical atoms there is another complication. To illustrate this, we consider *dissociation equilibrium* in a gas. Let the molecule be of the type $M=(C_1)_{p_1}(C_2)_{p_2}$; that is, let it be composed of p_1 atoms of type C_1 and p_2 atoms of type C_2. Let there be N molecules of type M, N_1 free atoms of type C_1, and N_2 free atoms of type C_2. Let C_1 be the total number of atoms of type C_1, and C_2 be the total number of atoms of type C_2; C_1 and C_2 are fixed numbers:

$$N_1 + p_1 N = C_1\,,$$

$$N_2 + p_2 N = C_2\,,$$

$$\exp[-F/kT] = \exp[-F_M/kT]\exp[-F_1/kT]\exp[-F_2/kT]P\,.$$

The factor $1/h^f$ is to be thought of as already incorporated into $\exp[-F/kT]$. The combinatorial factor P is equal to the number of different systems that can be formed holding N_1, N_2, and N fixed:

$$P = \frac{\text{number of permutations of identical atoms}}{\text{number of internal permutations}}$$

$$= \frac{C_1!\, C_2!}{N!\, N_1!\, N_2!\, \sigma^N}.$$

The σ which appears in the denominator is called the *symmetry number*. It is the number of permutations of identical atoms which can be generated by rotating a molecule. In other words, it is the number of equivalent orientations of a molecule or the *order of the symmetry group of M*. For example,

$$\begin{aligned} \mathrm{HCl}&: \quad \sigma = 1, \\ \mathrm{H_2}&: \quad \sigma = 2, \\ \mathrm{CH_4}&: \quad \sigma = 12. \end{aligned}$$

Letting $i = 1$ and 2, and defining

$$\exp[-F^*/kT] \equiv \frac{1}{C_1!\, C_2!} \exp[-F/kT],$$

$$\exp[-F_i^*/kT] \equiv \frac{1}{N_i!} \exp[-F_i/kT],$$

and

$$\exp[-F_M^*/kT] \equiv \frac{1}{N!\, \sigma^N} \exp[-F_M/kT],$$

we obtain

$$\exp[-F^*/kT] = \exp[-F_M^*/kT] \exp[-F_1^*/kT] \exp[-F_2^*/kT].$$

Applying this to our problem of the vapor pressure, we obtain the same μ for the polyatomic gases as for a monatomic gas, except for the following two modifications:

1. $\frac{5}{2} \log T \to (C_p/R) \log T$, where

$$\frac{C_p}{R} = 1 + \tfrac{1}{2} f \begin{cases} = \frac{5}{2} & \text{(monatomic)}, \\ = \frac{7}{2} & \text{(diatomic)}, \\ = 4 & \text{(triatomic)}. \end{cases}$$

2. The quantity i is different:

$$\mu = kT\{\log p - (1+\tfrac{1}{2}f)\cdot\log T - i\} + E_0,$$

monatomic: $$i = \log\left(\frac{(2\pi m)^{\frac{3}{2}}k^{\frac{5}{2}}}{h^3}\right),$$

diatomic: $$i = \log\left(\frac{(2\pi m)^{\frac{3}{2}}k^{\frac{5}{2}}}{h^3}\frac{8\pi A k}{\sigma h^2}\right),$$

tri- and poly-atomic: $$i = \log\left(\frac{(2\pi m)^{\frac{3}{2}}k^{\frac{5}{2}}}{h^3}\frac{8\pi^2(2\pi k)^{\frac{3}{2}}(A_1A_2A_3)^{\frac{1}{2}}}{\sigma h^3}\right).$$

a. Applications

1. *Solid-gas vapor pressure curve*: Using

$$\mu_{\text{solid}} = E_0 + kT\frac{9}{\nu_D^3}\int_0^{\nu_D}\nu^2\,d\nu\log\left(1-\exp\left[-\frac{h\nu}{kT}\right]\right)$$

along with the initial values, we can calculate the curve.

2. *Chemical equilibrium in a gas*:

$$W^*(N_1, N_2, N) = \text{constant}\times\exp\left[-\frac{1}{kT}[F^*(N) + F_1^*(N_1) + F_2^*(N_2)]\right].$$

For the most probable distribution, we obtain

$$\delta\log W^* = 0 \quad \text{for } \delta N_1 = -p_1\delta N \text{ and } \delta N_2 = -p_2\delta N.$$

Thus,

$$\mu_1\delta N_1 + \mu_2\delta N_2 + \mu_M\delta N = 0$$

or

$$\mu_M - p_1\mu_1 - p_2\mu_2 = 0,$$

which is the result obtained with thermodynamics. As a consequence, we now know the constants that appear in μ. Furthermore, it is also possible to calculate fluctuations in this manner.

b. Quantum theory and Nernst's theorem

Nernst's theorem, which says that all bodies at absolute zero have the same entropy (normalizable to zero), implies something about the density of quantum states in phase space. We have

$$F = -kT\log Z\,, \qquad \text{and} \qquad Z = \sum_n g_n \exp[-E_n/kT]\,,$$

where a degeneracy g_n for state n has been assumed. If the system is finite, then there are only discrete energy eigenvalues. Let E_0 be the lowest eigenvalue and E_1 be the second lowest. For $kT \ll E_1 - E_0$, we can make the following calculation:

$$Z = \exp[-E_0/kT]g_0\left(1 + \frac{g_1}{g_0}\exp\left[-\frac{E_1 - E_0}{kT}\right] + \ldots\right),$$

$$\log Z = -\frac{E_0}{kT} + \log g_0 + \frac{g_1}{g_0}\exp\left[-\frac{E_1 - E_0}{kT}\right] + \ldots,$$

$$F = E_0 - kT\log g_0 - kT\frac{g_1}{g_0}\exp\left[-\frac{E_1 - E_0}{kT}\right],$$

$$S = -\frac{\partial F}{\partial T} = k\log g_0 + \left(k + \frac{E_1 - E_0}{T}\right)\frac{g_1}{g_0}\exp\left[-\frac{E_1 - E_0}{kT}\right].$$

Letting $T \to 0$, we obtain

$$S = k\log g_0\,.$$

Therefore, Nernst's theorem requires $g_0 = 1$. This means

1. The lowest state is not degenerate;
2. The lowest state is sufficiently separated from the next-to-lowest state.

23. EINSTEIN'S THEORY OF BLACKBODY RADIATION

Shortcomings of the method of oscillators are

1. $E_\nu = (c^3/8\pi\nu^2)\varrho_\nu$ is a classical formula;
2. Molecules are not oscillators.

We now assume an arbitrary system with energy eigenvalues $E_1, E_2, \ldots$, but we do not specify either selection rules or degeneracies. On the other hand, we use the Bohr frequency condition:

$$E_n - E_m = h\nu_{nm} .$$

Einstein[5] introduced *transition probabilities* which describe the behavior of the system *statistically*:

$$(E_n > E_m) \begin{cases} \text{spontaneous emission} & n \to m: \quad A_m^n, \\ \text{induced emission} & n \to m: \quad B_m^n \varrho_\nu, \\ \text{absorption} & m \to n: \quad B_n^m \varrho_\nu . \end{cases}$$

The number of transitions in a time interval $\mathrm{d}t$ is given by

$$n \to m \qquad (A_m^n + B_m^n \varrho_\nu) N_n \mathrm{d}t = \mathrm{d}Z_{n\to m} ,$$

$$m \to n \qquad B_n^m \varrho_\nu N_m \mathrm{d}t = \mathrm{d}Z_{m\to n} .$$

Now assume thermal equilibrium:

$$N_n = g_n C \exp[-E_n/kT],$$

$$\mathrm{d}Z_{n\to m} = \mathrm{d}Z_{m\to n} .$$

Therefore,

$$g_n(A_m^n + B_m^n \varrho_\nu) \exp[-E_n/kT] = g_m B_n^m \varrho_\nu \exp[-E_m/kT] ,$$

$$A_m^n = \varrho_\nu \left(- B_m^n + \frac{g_m}{g_n} B_n^m \exp[(E_n - E_m)/kT] \right) .$$

Substituting $(E_n - E_m)/h = \nu$, we obtain

$$\varrho_\nu = \frac{A_m^n/B_m^n}{(g_m B_n^m / g_n B_m^n) \exp[h\nu/kT] - 1} .$$

If, for high temperatures, we require that this reduce to the Rayleigh-Jeans law (see p. 71), then

$$g_m B_n^m = g_n B_m^n , \tag{1}$$

$$\frac{A_m^n}{B_m^n} = \frac{8\pi h \nu^3}{c^3} . \tag{2}$$

[5] *Physik. Zeitschr.* **18**, 121 (1917).

For arbitrary temperatures, Planck's law follows from this immediately. The problem of calculating the A^n_m, B^n_m, and B^m_n separately can only be solved with quantum mechanics.

Momentum transfer between radiation and matter

We consider the Brownian motion that a particle undergoes in the radiation field. From the results of Chapter 3, we have

$$G_x = \int_0^\tau X\,dt\,, \qquad \overline{G_x} = 0\,, \qquad \overline{G_x^2} = 2kTW\tau\,.$$

1. *Calculation of the resistive force W acting on a particle moving slowly in the radiation field.* Denote the rest system of the radiation by K and the rest system of the particle by K'. During absorption, the recoil momentum is, for the transition $n \to m$ in K,

$$\frac{h\nu}{c}\cos\theta = \frac{E_n - E_m}{c}\cos\theta\,,$$

where θ is the angle between the photon and particle momenta; during emission, the recoil momentum is

$$-\frac{h\nu}{c}\cos\theta\,.$$

Define S by

$$S = g_n \exp[-E_n/kT] + g_m \exp[-E_m/kT] + \dots\,.$$

The fraction of time which the particle spent in state n is

$$g_n \exp[-E_n/kT]/S\,;$$

the fraction of time spent in state m is

$$g_m \exp[-E_m/kT]/S\,.$$

The numbers of absorptions and emissions into solid angle

$d\Omega'$ are, respectively,

$$\frac{1}{S} g_m \exp[-E_m/kT] B_n^m \varrho_\nu' \frac{d\Omega'}{4\pi}$$

and

$$\frac{1}{S} g_n \exp[-E_n/kT] B_m^n \varrho_\nu' \frac{d\Omega'}{4\pi} .$$

(Note that the prime refers to the system K' and that ϱ_ν' is not isotropic.) Let $\nu_0 = (E_n - E_m)/h$. Then,

$$-Wv = \frac{h\nu_0}{cS} g_n B_m^n (\exp[-E_m/kT] - \exp[-E_n/kT]) \times \int \varrho_\nu'(\theta') \cos\theta' \frac{d\Omega'}{4\pi} .$$

Without proof, we state that the following relation holds [A-7]:

$$\frac{\varrho_{\nu'}'}{\nu'^3} = \frac{\varrho_\nu}{\nu^3} .$$

Further, to first order in v/c, we have a Doppler shift given by

$$\nu = \nu' \left(1 + \frac{v}{c} \cos\theta'\right) .$$

Therefore,

$$\varrho_{\nu'}' = \left(1 + \frac{v}{c} \cos\theta'\right)^{-3} \varrho_{[1+(v/c)\cos\theta']\nu'}$$

$$\varrho_{\nu_0}' \cong \left(1 - 3\frac{v}{c}\cos\theta'\right)\left[\varrho_{\nu_0} + \left(\frac{\partial\varrho_\nu}{\partial\nu}\right)_0 \nu_0 \frac{v}{c}\cos\theta'\right]$$

$$\cong \varrho_{\nu_0} + \frac{v}{c}\cos\theta'\left[\nu_0\left(\frac{\partial\varrho_\nu}{\partial\nu}\right)_0 - 3\varrho_{\nu_0}\right],$$

$$\int \varrho_{\nu_0}' \cos\theta' \frac{d\Omega'}{4\pi} = -\frac{v}{c}\left\{\varrho_{\nu_0} - \frac{1}{3}\nu_0\left(\frac{\partial\varrho_\nu}{\partial\nu}\right)_0\right\}$$

$$= -\frac{v}{c}\left\{\left(-\frac{\nu^4}{3}\right)\frac{\partial}{\partial\nu}\left(\frac{\varrho_\nu}{\nu^3}\right)\right\}_{\nu=\nu_0} .$$

Then, again writing ν instead of ν_0, and interchanging n

and m, we obtain

$$W = \frac{h\nu}{c^2} \frac{1}{S} g_m B_n^m \exp[-E_m/kT](1 - \exp[-h\nu/kT]) \times \left[-\frac{\nu^4}{3} \frac{\partial}{\partial\nu}\left(\frac{\varrho_\nu}{\nu^3}\right)\right].$$

In equilibrium, we have the Planck formula,

$$\varrho_\nu = \frac{8\pi h}{c^3} \frac{\nu^3}{\exp[h\nu/kT]-1};$$

$$-\frac{\partial}{\partial\nu}\left(\frac{\varrho_\nu}{\nu^3}\right) = \frac{8\pi h}{c^3} \frac{\exp[h\nu/kT]}{(\exp[h\nu/kT]-1)^2} \frac{h}{kT}$$

$$= \frac{1}{\nu^3} \varrho_\nu \frac{h}{kT} \cdot \frac{\exp[h\nu/kT]}{\exp[h\nu/kT]-1}$$

$$W = \frac{1}{3}\left(\frac{h\nu}{c}\right)^2 \frac{1}{kT} \frac{g_m}{S} \exp[-E_n/kT] B_n^m \varrho_\nu,$$

$(g_m/S)\exp[-E_n/kT]B_n^m\varrho_\nu$ = number of absorptions per second = $\frac{1}{2}Z$ (Z = total number of *all* processes per second). Hence

$$W = \frac{1}{3}\left(\frac{h\nu}{c}\right)^2 \frac{1}{2kT} Z.$$

2. *Calculation of* $\overline{G^2}$.

$$\overline{G^2} = 2kTW\tau = \frac{1}{3}\left(\frac{h\nu}{c}\right)^2 Z\tau,$$

where $Z\tau$ = total number of *all* transitions (induced and spontaneous emissions plus absorptions) during a time τ.

Interpretation. Each transition, including spontaneous emission, imparts a recoil momentum $h\nu/c$ to the atom:

$$\overline{G^2} = \left(\frac{h\nu}{c}\right)^2 \overline{\cos^2\theta}\, Z\tau = \frac{1}{3}\left(\frac{h\nu}{c}\right)^2 Z\tau.$$

Thus, we are compelled to ascribe a recoil action even to the spontaneous emission process. This is in contradiction to interference observations, from which it is known that spherical waves can be coherent.

This paradox is resolved by the remark that an exact knowledge of the momentum of the particle is required for a measurement of the recoil momentum. According to the Heisenberg uncertainty principle,[6] exact knowledge of the momentum would exclude the exact knowledge of position that would be necessary for observation of interference.

24. QUANTUM STATISTICS OF IDENTICAL PARTICLES

In the case of radiation, the state was completely determined by giving the energy $n_s h\nu_s$ of each normal mode. In particle language, this means giving the number of photons n_s in state s. One can try to use the same description in the case of an ideal gas. Let the word "state" denote the state of an atom and the word "cell" denote the state of the entire system. According to our previous considerations (see Sec. 7), for a specified system (n_s), there are

$$P = \frac{n!}{n_1!n_2!\ldots n_s!\ldots}$$

distinct cells.

If we consider radiation, then, because of the wave nature of radiation, the situation is different. Guided by the wave nature of matter, we can tentatively try the same procedure with gases that we used for radiation.

For low densities, such that $n_s \to 0$ or 1 for all s, we have $P \to N!$. The density of states is then

$$\frac{1}{N!h^{3N}} \quad \text{instead of} \quad \frac{1}{h^{3N}}.$$

In the particle picture, this has the consequence that the particles are no longer statistically independent. For example, consider a case for which there are two states, A and B, and two particles.

[6] See W. PAULI, *Wave Mechanics*.

		Statistical Weight		
n_A	n_B	B.E.	statistical independence	F.D.
0	2	1	1	0
1	1	1	2	1
2	0	1	1	0

In the case considered here, the "*statistics of symmetrical states*" or *Bose-Einstein statistics* (B.E.), the particles have the tendency to condense into groups.

In addition, there is the "*statistics of antisymmetric states*" or *Fermi-Dirac statistics* (F.D.), for which the n_s are restricted to have only the values 0 and 1. In this case, it is evident that the particles repel one another.

In the case of radiation, the total number of particles is not specified. In order to preserve the analogy with radiation as nearly as possible, we start from the grand canonical ensemble (see Sec. 14). Then we do not have to write the $N!$ because the particles are now considered indistinguishable.

$$W^*(n_1, n_2, \ldots, n_s, \ldots) = \exp\left[\alpha\{\Omega + \mu N - E(n_1, \ldots, n_s, \ldots)\}\right].$$

For *radiation*, the μ in this formula must be set equal to zero, since N is not an independent variable.

We now consider an *ideal gas*. Let ε_s = the energy in cell s. Then,

$$E = \sum_s n_s \varepsilon_s,$$

$$N = \sum_s n_s,$$

and

$$W^* = \exp\left[\alpha\Omega\right] \prod_s \exp\left[\alpha(\mu - \varepsilon_s) n_s\right].$$

From $\sum W^* = 1$ we obtain

B.E.: $n_s = 0, \ldots, \infty$: $$1 = \exp[\alpha\Omega] \prod_s \frac{1}{1 - \exp[\alpha(\mu - \varepsilon_s)]},$$

F.D.: $n_s = 0, 1$: $$1 = \exp[\alpha\Omega] \prod_s \{1 + \exp[\alpha(\mu - \varepsilon_s)]\}.$$

Therefore,

$$\text{B.E.:} \qquad \alpha\Omega = \sum_s \log\left\{1 - \exp[\alpha(\mu - \varepsilon_s)]\right\},$$

$$\text{F.D.:} \qquad \alpha\Omega = -\sum_s \log\left\{1 + \exp[\alpha(\mu - \varepsilon_s)]\right\}.$$

From $N = -\partial\Omega/\partial\mu$ we obtain

$$\text{B.E.:} \qquad N = \sum_s \frac{1}{\exp[-\alpha(\mu - \varepsilon_s)] - 1},$$

$$\text{F.D.:} \qquad N = \sum_s \frac{1}{\exp[-\alpha(\mu - \varepsilon_s)] + 1}.$$

From the requirement $N > 0$, we see

$$\text{B.E.:} \qquad -\infty \leqslant \alpha\mu \leqslant 0,$$

$$\text{F.D.:} \qquad -\infty < \alpha\mu < +\infty.$$

From $S = -(\partial\Omega/\partial T)_\mu$ we obtain

$$\left.\begin{array}{l}\text{B.E.:}\\ \text{F.D.:}\end{array}\right\} \qquad S = -\frac{\Omega}{T} - \frac{1}{T}\sum_s \frac{\exp[\alpha(\mu - \varepsilon_s)]}{1 \mp \exp[\alpha(\mu - \varepsilon_s)]}(\mu - \varepsilon_s),$$

$$\left.\begin{array}{l}\text{B.E.}\\ \text{F.D.}\end{array}\right\} \qquad S = \frac{-F + E}{T} = \frac{-\Omega - \mu N + E}{T},$$

where

$$E = \sum_s \frac{\varepsilon_s}{\exp[-\alpha(\mu - \varepsilon_s)] \mp 1}.$$

a. Transition to integrals

For an ideal gas of mass points the density of states is

$$\frac{dZ}{d\varepsilon} = V\frac{2\pi(2m)^{\frac{3}{2}}}{h^3}\varepsilon^{\frac{1}{2}}.$$

With

$$\frac{\varepsilon}{kT} = \alpha\varepsilon \equiv x, \qquad d\varepsilon = kT\,dx, \qquad \text{and} \qquad A = \alpha\mu = \frac{\mu}{kT}$$

we obtain[7]

$$\Omega = \pm V(kT)^{\frac{5}{2}} \frac{2\pi (2m)^{\frac{3}{2}}}{h^3} \int_0^\infty \log(1 \mp \exp[A - x])\sqrt{x}\,dx\,.$$

With $p = -(\partial\Omega/\partial V)_{\mu,T}$, we obtain $\Omega = -pV$, as it should be. Integrating by parts, we obtain

$$\pm \int_0^\infty \log(1 \mp \exp[A - x])\,\tfrac{2}{3}\mathrm{d}(x^{\frac{3}{2}})$$

$$= \pm\,\tfrac{2}{3}x^{\frac{3}{2}} \log(1 \mp \exp[A - x])\Big|_0^\infty - \frac{2}{3}\int_0^\infty x^{\frac{3}{2}} \frac{\exp[A - x]\,dx}{1 \mp \exp[A - x]},$$

and hence[7]

$$-\Omega = pV = V(kT)^{\frac{5}{2}} \frac{2\pi (2m)^{\frac{3}{2}}}{h^3} \frac{2}{3} \int_0^\infty \frac{x^{\frac{3}{2}}\,dx}{\exp[-A + x] \mp 1}\,.$$

Further,

$$N = V \frac{2\pi (2m)^{\frac{3}{2}}}{h^3} (kT)^{\frac{3}{2}} \int_0^\infty \frac{\sqrt{x}\,dx}{\exp[-A + x] \mp 1}\,,$$

$$E = V \frac{2\pi (2m)^{\frac{3}{2}}}{h^3} (kT)^{\frac{5}{2}} \int_0^\infty \frac{x^{\frac{3}{2}}\,dx}{\exp[-A + x] \mp 1}\,.$$

Therefore,

$$-\Omega = pV = 2E/3,$$

which is the result for an ideal gas.

With the definitions

$$F_{\mp}(A) \equiv \frac{2}{\sqrt{\pi}} \int_0^\infty \frac{\sqrt{x}\,dx}{\exp[-A + x] \mp 1}$$

and

$$G_{\mp}(A) = \frac{4}{3\sqrt{\pi}} \int_0^\infty \frac{x^{\frac{3}{2}}\,dx}{\exp[-A + x] \mp 1}$$

[7] For Bose-Einstein statistics, the upper sign is to be taken; for Fermi-Dirac statistics, the lower sign is to be taken.

we obtain

$$N = V \frac{(2m\pi kT)^{\frac{3}{2}}}{h^3} F_{\mp}(A),$$

$$E = V \frac{3}{2} \frac{(2\pi m)^{\frac{3}{2}} (kT)^{\frac{5}{2}}}{h^3} G_{\mp}(A) .$$

Here, $A = \mu/kT$ is to be thought of as a parameter which is to be eliminated from the two equations. By partial integration we find

$$G'_{\mp}(A) = F_{\mp}(A) .$$

b. Limiting cases

1. *Dilute F.D. or B.E. gas* ($A \to -\infty$).

$$\frac{1}{\exp[-A+x] \mp 1} = \frac{\exp[-|A|-x]}{1 \mp \exp[-|A|-x]}$$

$$= \sum_{n=1}^{\infty} (\pm 1)^{n-1} \exp[(-|A|-x)n], \quad (A<0).$$

With

$$\frac{2}{\sqrt{\pi}} \int_0^{\infty} \exp[-nx] x^{\frac{1}{2}} \mathrm{d}x = \frac{1}{n^{\frac{3}{2}}}$$

and

$$\frac{4}{3\sqrt{\pi}} \int_0^{\infty} \exp[-nx] x^{\frac{3}{2}} \mathrm{d}x = \frac{1}{n^{\frac{5}{2}}},$$

we obtain, for $A < 0$, in the B.E. case,

$$F_{-}(A) = \sum_1^{\infty} \frac{\exp[-|A|n]}{n^{\frac{3}{2}}},$$

$$G_{-}(A) = \sum_1^{\infty} \frac{\exp[-|A|n]}{n^{\frac{5}{2}}},$$

and in the F.D. case

$$F_{+}(A) = \sum_1^{\infty} (-1)^{n-1} \frac{\exp[-|A|n]}{n^{\frac{3}{2}}},$$

$$G_{+}(A) = \sum_1^{\infty} (-1)^{n-1} \frac{\exp[-|A|n]}{n^{\frac{5}{2}}}.$$

For $|A| \gg 1$, only the first term in the sums need be considered:

$$N = V \frac{(2m\pi kT)^{\frac{3}{2}}}{h^3} e^A ,$$

$$A = \frac{\mu}{kT} = -\log\left[\frac{V}{N}\frac{(2m\pi kT)^{\frac{3}{2}}}{h^3}\right].$$

This is the μ of a normal monatomic gas. Further, from

$$G_- = F_-$$

follows

$$E = \tfrac{3}{2}NkT , \qquad -\Omega = pV = NkT .$$

To a higher approximation we obtain per mole

$$pV = RT\left(1 \mp 2^{-\frac{5}{2}} h^3 \frac{L}{V}(2\pi mkT)^{-\frac{3}{2}} + \ldots\right).$$

2. *F.D. gas at low temperature* ($A \to \infty$). For $F_+(A)$ and $G_+(A)$ there are the following asymptotic series:

$$F_+(A) = \frac{4}{3\sqrt{\pi}} A^{\frac{3}{2}}\left(1 + \frac{\pi^2}{8A^2} + \ldots\right),$$

$$G_+(A) = \frac{8}{15\sqrt{\pi}} A^{\frac{5}{2}}\left(1 + \frac{5\pi^2}{8A^2} + \ldots\right).$$

Proof:

Let

$$J = \int_0^\infty \frac{f(x)\,\mathrm{d}x}{\exp[-A + x] + 1}, \qquad x = A + y ,$$

$$f(A + y) = g(y) , \qquad \text{and} \qquad \int_{-A}^{y} g(y)\,\mathrm{d}y = G(y) .$$

Then,

$$J = \int_{-A}^{\infty} \frac{g(y)\,\mathrm{d}y}{e^y + 1} = \int_{-A}^{\infty} \frac{G'(y)\,\mathrm{d}y}{e^y + 1} = \underbrace{\frac{G(y)}{e^y + 1}\Bigg|_{-A}^{\infty}}_{0} + \int_{-A}^{\infty} \frac{G(y)\,e^y\,\mathrm{d}y}{(e^y + 1)^2} .$$

The Taylor series for G (with the remainder) is

$$G(y) = G(0) + yG'(0) + \tfrac{1}{2}y^2G''(0) + \dots .$$

Hence

$$\int_{-A}^{\infty} \frac{G(y)\,e^y}{(e^y+1)^2}\,dy \simeq G(0)\int_{-\infty}^{+\infty} \frac{e^y}{(e^y+1)^2}\,dy$$

$$+ G'(0)\int_{-\infty}^{+\infty} \frac{y\,e^y\,dy}{(e^y+1)^2} + \tfrac{1}{2}G''(0)\int_{-\infty}^{+\infty} \frac{y^2\,e^y\,dy}{(e^y+1)^2} + \dots ,$$

$$J = G(0) + G''(0)\,\frac{\pi^2}{6} + \dots ,$$

$$G(0) = \int_{-A}^{0} g(y)\,dy = \int^{A} f(x)\,dx, \quad G'(0) = f(A), \quad G''(0) = f'(A), \dots ,$$

$$J = \int_0^{\infty} \frac{f(x)\,dx}{\exp[-A+x]+1} = \int_0^{A} f(x)\,dx + \frac{\pi^2}{6}f'(A) + \dots .$$

From this the assertion follows immediately.

Now, let $f(x) = f(\varepsilon/kT) = \varphi(\varepsilon)$, and let $A = \mu/kT$. Then we have

$$\int_0^{\infty} \frac{\varphi(\varepsilon)\,d\varepsilon}{\exp[(-\mu+\varepsilon)/kT]+1} \simeq \int_0^{\mu} \varphi(\varepsilon)\,d\varepsilon + \frac{\pi^2}{6}(kT)^2\varphi'(\mu) + \dots .$$

In the limit as $T \to 0$, the integral on the left can be replaced by the first term on the right, which is an integral whose range only extends to μ. The function φ is given by $\varphi(\varepsilon) \sim \varepsilon^{\frac{1}{2}}$ for N, and $\sim \varepsilon^{\frac{3}{2}}$ for E.

Zeroth-order approximation:

$$N = V\,\frac{2\pi(2m)^{\frac{3}{2}}}{h^3}\,\frac{2}{3}\,\mu_0^{\frac{3}{2}}, \qquad E_0 = V\,\frac{2\pi(2m)^{\frac{3}{2}}}{h^3}\,\frac{2}{5}\,\mu_0^{\frac{5}{2}}.$$

The zero-point energy becomes

$$E_0 = N\,\frac{3}{40}\left(\frac{6}{\pi}\right)^{\frac{2}{3}}\frac{h^2}{m}\left(\frac{N}{V}\right)^{\frac{2}{3}}.$$

First-order approximation:

$$N = V \frac{2\pi (2m)^{\frac{3}{2}}}{h^3} \frac{2}{3} \mu^{\frac{3}{2}} \left[1 + \frac{\pi^2}{8\mu^2} (kT)^2\right],$$

$$E = V \frac{2\pi (2m)^{\frac{3}{2}}}{h^3} \frac{2}{5} \mu^{\frac{5}{2}} \left[1 + \frac{5\pi^2}{8\mu^2} (kT)^2\right]$$

$$= N \frac{3}{5} \mu \left\{1 + \frac{\pi^2}{2\mu^2} (kT)^2\right\}.$$

If we keep the definition of μ_0 used in the zeroth-order approximation, then we obtain

$$\mu^{\frac{3}{2}} \left[1 + \frac{\pi^2}{8\mu^2} (kT)^2\right] = \mu_0^{\frac{3}{2}}$$

or

$$\mu = \mu_0 \left[1 - \frac{\pi^2}{12\mu_0^2} (kT)^2 + \dots\right],$$

$$E = N \frac{3}{5} \mu_0 \left[1 + \frac{5\pi^2}{12\mu_0^2} (kT)^2 + \dots\right],$$

$$C_v = \left(\frac{\partial E}{\partial T}\right)_N N \frac{\pi^2}{2\mu_0} k^2 T .$$

For the entropy we have

$$S = \frac{E - F}{T} = \frac{E - \mu N + pV}{T} = \frac{\frac{5}{2}E - \mu N}{T}$$

$$= N \frac{\pi^2}{2\mu_0} k^2 T \qquad \text{(Nernst's theorem)} .$$

Application to an electron gas: In order to take account of the electron spin, we can consider the gas as a mixture of a two-component gas with opposite spin directions. In the absence of external forces, each gas contains $N/2$ particles. Therefore, the previous formulas are valid if we replace N by $N/2$ and E by $E/2$. For $\mu_0 \gg kT$, we obtain

$$E_0 = N \tfrac{3}{5} \mu_0 ,$$

$$N = 2V \frac{2\pi (2m)^{\frac{3}{2}}}{h^3} \frac{2}{3} \mu_0^{\frac{3}{2}} .$$

In the presence of an external magnetic field, there is an additional energy $\mu_B \boldsymbol{H}$, where $\mu_B = e\hbar/(2mc)$ is the Bohr magneton. Let us denote the cases of spin parallel to $\boldsymbol{H}$ and antiparallel to $\boldsymbol{H}$ by the subscripts 1 and 2, respectively. The equilibrium condition is

$$\mu_1 - \mu_B H = \mu_2 + \mu_B H .$$

This equation and the equation

$$N = V \frac{2\pi (2m)^{\frac{3}{2}}}{h^3} \left(\tfrac{2}{3}\mu_1^{\frac{3}{2}} + \tfrac{2}{3}\mu_2^{\frac{3}{2}}\right)$$

together determine μ_1 and μ_2. Further, we have

$$\overline{M} = \mu_B(N_1 - N_2) = \mu_B V \frac{2\pi (2m)^{\frac{3}{2}}}{h^3} \tfrac{2}{3} \left(\mu_1^{\frac{3}{2}} - \mu_2^{\frac{3}{2}}\right) .$$

For $\mu_B H \ll \mu_0$, we have

$$\mu_1^{\frac{3}{2}} - \mu_2^{\frac{3}{2}} \simeq \tfrac{3}{2}\mu_0^{\frac{1}{2}}(\mu_1 - \mu_2) ,$$

and

$$\overline{M} = 2H\mu_B^2 V \frac{2\pi (2m)^{\frac{3}{2}}}{h^3} \mu_0^{\frac{1}{2}} .$$

This is a weak, temperature-independent paramagnetism. (The result is only qualitative, since forces between the electrons have been neglected.) Superimposed on this paramagnetism is an orbital diamagnetism which is $\frac{1}{3}$ as strong [A-8].

3. *B.E. gas. Degeneracy.* The series given for the first limiting case are always convergent for the B.E. gas. Particularly, for $A = 0$, the values are $F(0) = 2.615$ and $G(0) = 1.34$. Hence

$$N_0 = \frac{V}{h^3} (2\pi mkT)^{\frac{3}{2}} \cdot 2.615 = N_{\max} ,$$

$$E_0 = \tfrac{3}{2} N_0 kT \frac{1.34}{2.615} ,$$

$$S_0 = \tfrac{5}{2} kN_0 \frac{1.34}{2.615} .$$

That is, for fixed V and T, there is a maximum number of particles that can be in the gas. If we try to add more, then, following Einstein, what happens is the following: The particles go into the condensed state ($\varepsilon=0$, $S=0$, $p=0$, $E=0$, and, therefore, $\mu=0$) and, in a sense, form a second phase which is in equilibrium with the first phase, since $\mu=0$ for both.

One can object that replacement of the sum by an integral is no longer correct for $A=0$. If the calculation is carried out with sums, then this bounded N_0 is not obtained since for $\mu=0$ the function is singular at $\varepsilon=0$:

$$\frac{\sqrt{\varepsilon}}{\exp[\varepsilon/kT]-1}=O\left(\frac{1}{\sqrt{\varepsilon}}\right).$$

The result depends on the type of vessel. If the vessel is sufficiently large, then the above theory is nevertheless essentially correct.

Bibliography

Chapter 1

L. Boltzmann, *Vorlesungen über Gastheorie* (Verlag von Johann Ambrosius Barth, Leipzig, 1895).

P. Ehrenfest and T. Ehrenfest, *Begriffliche Grundlagen der statistischen Auffasung in der Mechanik*, article in *Encyklopädie d. mathematischen Wissenschaften*, IV 2, II, Heft 6 (B. G. Teubner, Leipzig, 1912); reprinted in *Collected Scientific Papers* (Interscience, New York, 1959); translation: *The Conceptual Foundations of the Statistical Approach in Mechanics* (Cornell University Press, Ithaca, 1959).

S. Chapman and T. G. Cowling, *Mathematical Theory of Non-Uniform Gases* (University Press, Cambridge, 1952).

Chapter 2

J. W. Gibbs, *Elementary Principles in Statistical Mechanics Developed with Especial Reference to the Rational Foundation of Thermodynamics* (Scribner, New York, 1902; reprinted: Dover, New York, 1960).

A. Einstein, Eine Theorie der Grundlagen der Thermodynamik, *Ann. Physik* **11**, 170 (1903).

Chapter 3

G. L. de Haas and H. A. Lorentz, *Die Brown'sche Bewegung und einige verwandte Erscheinungen* (Vieweg, Braunschweig, 1913).

Chapter 4

M. Planck, *The Theory of Heat Radiation*, English translation (Blakiston's, Philadelphia, 1914).

General References

R. H. Fowler, *Statistical Mechanics* (University Press, Cambridge, 1955).

D. ter Haar, *Elements of Statistical Mechanics* (Rinehart, New York, 1954).

K. Huang, *Statistical Mechanics* (Wiley, New York, 1963).

P. Jordan, *Statistische Mechanik auf quantentheoretischer Grundlage* (Vieweg, Braunschweig, 1933).

L. D. Landau and E. M. Lifshitz, *Statistical Physics*, English translation (Addison-Wesley, Reading, 1958).

R. C. Tolman, *The Principles of Statistical Mechanics* (Oxford University Press, London, 1938).

Appendix. Comments by the Editor

[A-1] (pp. 12, 14, 15, 17). The *Boltzmann equation* is usually written in the form

$$\frac{\mathrm{d}f}{\mathrm{d}t} = \left(\frac{\partial f}{\partial t}\right)_{\text{coll}}$$

where

$$\frac{\mathrm{d}f}{\mathrm{d}t} = \frac{\partial f}{\partial t} + \frac{\partial f}{\partial \boldsymbol{x}} \cdot \boldsymbol{v} + \frac{\partial f}{\partial \boldsymbol{v}} \cdot \boldsymbol{K}/m$$

is the *kinematic term* and

$$\left(\frac{\partial f}{\partial t}\right)_{\text{coll}} = \int\int \mathrm{d}^3 V \, \mathrm{d}^2 \lambda [f(\boldsymbol{v}')f(\boldsymbol{V}') - f(\boldsymbol{v})f(\boldsymbol{V})] wq$$

is the *collision term*. There exist two distinct situations of importance where the collision term is negligible.

1. *Local equilibrium*: $(\partial f/\partial t)_{\text{coll}} = 0$. By definition a local equilibrium distribution annihilates the collision term. This implies that Eq. [4.4] on p. 10 must hold. Now this equation as well as Eqs. [4.4a], [4.7a], [4.7b] are not changed by letting f depend on $\boldsymbol{x}$ and t and by including a potential energy E_{pot}. Therefore the most general form of a *local equilibrium distribution* is

$$f_{\mathrm{L}}(\boldsymbol{v}, \boldsymbol{x}, t) = A \exp\left\{- \beta(\boldsymbol{x}, t) \left[\frac{m}{2} \boldsymbol{v}^2 + E_{\text{pot}}(\boldsymbol{x}, t) - m\boldsymbol{v} \cdot \boldsymbol{c}(\boldsymbol{x}, t)\right]\right\},$$

where any $(\boldsymbol{x}, t)$-dependence of A can be absorbed into E_{pot}

which therefore may also depend on time. The function $\beta^{-1}(\boldsymbol{x}, t) = kT(\boldsymbol{x}, t)$, where k is Boltzmann's constant, defines a *local temperature* $T(\boldsymbol{x}, t)$, and $\boldsymbol{c}(\boldsymbol{x}, t)$ is a *local drift*. The distribution f_{L} is restricted by the condition that the collision term describes the collisions properly which means that the frequencies involved are small compared to the collision frequencies. In addition, it has to satisfy the Boltzmann equation, which now reduces to the kinematic equation

$$\frac{\mathrm{d}f_{\mathrm{L}}}{\mathrm{d}t} = 0 \ .$$

A *stationary distribution* is a special case of an f_{L} because $\partial f/\partial t = 0$ implies $\mathrm{d}\mathscr{H}/\mathrm{d}t = 0$ and hence again $(\partial f/\partial t)_{\mathrm{coll}} = 0$. However, now the Boltzmann equation reads

$$\frac{\partial f}{\partial \boldsymbol{x}} \cdot \boldsymbol{v} + \frac{\partial f}{\partial \boldsymbol{v}} \cdot \boldsymbol{K}/m = 0 \ ,$$

which implies that E_{pot} is time-independent and $\beta = 2\alpha/m =$ $=$ constant; $\boldsymbol{c} = \boldsymbol{c}^0 + \boldsymbol{\omega} \times \boldsymbol{x}$ is perpendicular to $\boldsymbol{K} = -\,\partial E_{\mathrm{pot}}/\partial \boldsymbol{x}$ and $\boldsymbol{c}^0 =$ constant, $\boldsymbol{\omega} =$ constant.

2. *High-frequency limit*: $|\partial f/\partial t| \gg |(\partial f/\partial t)_{\mathrm{coll}}|$. The Boltzmann equation again reduces to $\mathrm{d}f/\mathrm{d}t = 0$. A high-frequency solution of this equation is called a *collisionless distribution*. This case is of importance for high-density Fermi liquids, in particular, electron plasmas. However, because of the strong interparticle forces, one has to include in the potential energy a distribution-dependent term of the form $\int \mathrm{d}^3v' \varphi(\boldsymbol{v}, \boldsymbol{v}') f(\boldsymbol{v}', \boldsymbol{x}, t)$, which gives rise to a so-called *Vlasov term* in the collisionless Boltzmann equation and to a *collective mode* (plasmon, zero sound).

An approximation of the collision term often used for transport processes (see Sec. 5) is the *collision time approximation*,

$$\left(\frac{\partial f}{\partial t}\right)_{\mathrm{coll}} = -\,(f - f_0)/\tau \ ,$$

where $f_0 = A \exp[-\alpha(\boldsymbol{v} - \boldsymbol{c})^2]$ is a stationary distribution. If $f - f_0$ oscillates as $\exp(-i\omega t)$ and if the system is perturbed by a weak external force $\boldsymbol{K} = \boldsymbol{K}_{\text{ext}}$, the solution of the Boltzmann equation becomes, to lowest order in $\boldsymbol{K}_{\text{ext}}$,

$$f = \left\{1 + \frac{\tau}{1 - i\omega\tau}\frac{2\alpha}{m}\boldsymbol{u}\cdot\boldsymbol{K}_{\text{ext}}\right\} f_0 ,$$

where $\boldsymbol{u} = \boldsymbol{v} - \boldsymbol{c}$. Note that a temperature gradient may be considered as an external force according to the substitution

$$\frac{2\alpha}{m}\boldsymbol{K}_{\text{ext}} \to \frac{m}{2}\boldsymbol{u}^2\frac{1}{kT^2}\frac{\partial T}{\partial \boldsymbol{x}}$$

with $2\alpha/m = 1/(kT)$. For $\omega = 0$ the above f then becomes a special case of the expression given on p. 17.

The frequency range $\omega\tau \ll 1$ is called the *hydrodynamic domain*; $\omega\tau \gg 1$ is the *collisionless domain.*

[A-2] (p. 29). With modern computers it has become possible to calculate such time averages and hence, in principle, to check the ergodic hypothesis "by experiment." See, for example, A. Rahman, *Phys. Rev.* **136**, A 405 (1964); L. Verlet, *Phys. Rev.* **159**, 98 (1967).

[A-3] (pp. 32, 39, 53, 54). The virial theorem

$$2N\overline{E_{\text{kin}}} - (V^{\text{ext}} + V^{\text{int}}) = N\frac{m}{2}\frac{\mathrm{d}^2}{\mathrm{d}t^2}\overline{\boldsymbol{q}^2} = 0$$

follows trivially from Newton's equation and the formula for the Brownian motion $\overline{\boldsymbol{q}^2} = 2Dt$ derived in Secs. 16, 17, 18. Here

$$V^{\text{ext}} = -\sum_i \boldsymbol{q}_i\cdot\boldsymbol{F}_i^{\text{ext}} = \oint p\boldsymbol{q}\cdot\mathrm{d}\boldsymbol{S} = 3pV$$

is the *external virial* due to the forces acting on the walls. For central two-body forces

$$\boldsymbol{F}_i^{\text{int}} = -\sum_{k\neq i}\frac{\boldsymbol{q}_i - \boldsymbol{q}_k}{r_{ik}}U'(r_{ik}); \qquad r_{ik} = |\boldsymbol{q}_i - \boldsymbol{q}_k| ,$$

and the *internal virial* is

$$V^{\text{int}} = -\sum_i \boldsymbol{q}_i \cdot \boldsymbol{F}_i^{\text{int}} = \tfrac{1}{2} \sum_{i,k\neq i} r_{ik}\, U'(r_{ik})$$

$$= \frac{N^2}{2}\frac{1}{V}\int_\sigma^\infty r\,U'(r)\,4\pi r^2\,\mathrm{d}r = \frac{3}{V}A(T)\,.$$

By partial integration one finds per mole, assuming that $U(r)$ tends to zero stronger than r^{-3} for $r \to \infty$,

$$\overline{A(T)} = -\frac{2\pi}{3}L^2\sigma^3\,\overline{U(\sigma)} - \frac{L^2}{2}\int_\sigma^\infty U(r)\,4\pi r^2\,\mathrm{d}r\,.$$

Now $U(\sigma) = 0$ by definition (see Fig. 13.3), but application of the virial and equipartition theorems (p. 32) to the radial motion gives

$$\overline{U(\sigma)} = -\overline{r\,U'(r)} = 2\overline{E_{\text{kin}}^{\text{rad}}} = kT\,.$$

Together with $\overline{E_{\text{kin}}} = \frac{3}{2}kT$ the expressions of p. 54 then readily follow.

[A-4] (p. 56). The functions F, F_1, and F_2 are the same because homogeneity implies that the Hamiltonians are sums over the respective particle numbers N, N_1, and N_2 of the same 1-, 2-, etc. particle Hamiltonians.

[A-5] (p. 57). This ceases to be the case in the vicinity of a *phase transition* where the kth Fourier component of the fluctuations, $\overline{(\Delta N_{1k})^2}$, is of the form

$$\overline{(\Delta N_{1k})^2} = \frac{\text{constant}}{(\xi^{-2} + k^2)^{1-\eta/2}}\,.$$

Here

$$\xi(T) = \frac{\text{constant}}{|T - T_c|^\nu}$$

is the *correlation length* and T_c the *transition temperature.* While classical theory (Ornstein-Zernike) yields for the *critical exponents* $\nu = 0.5$, $\eta = 0$, experimental evidence favors slightly larger values of these parameters. For more detail see, for example, H. E. Stanley, *Phase Transitions and Critical Phenomena* (Oxford University Press, New York, 1971).

[A-6] (p. 85). The meaning of the zero-point energy which Planck had introduced somewhat reluctantly in 1911 was a much discussed problem in the days of the old de Broglie–Sommerfeld quantum theory. Pauli discussed it extensively with Otto Stern in Hamburg in the early 1920s. Stern had calculated, but never published, the vapor pressure difference between the isotopes 20 and 22 of Neon. Without zero-point energy this difference would be large enough for easy separation of the isotopes, which is not the case in reality.

Pauli, on the other hand, had calculated, but also not published, the zero-point energy of radiation and came to the conclusion that its gravitational effect would be so large that the radius of the universe "could not even reach to the moon" (short wavelengths were cut off at the classical electron radius).

For more details on this question see C. P. Enz and A. Thellung, *Helv. Phys. Acta* **33**, 839 (1960).

[A-7] (pp. 85, 96). Let the adiabatic compression of the cavity of p. 85 be produced by a piston moving with velocity v. Then a photon of frequency ν incident on the piston under an angle θ will be reflected with frequency ν' determined by the Doppler shift formula

$$\nu' = \nu \left(1 + \frac{2v}{c} \cos\theta\right) .$$

The work done by the radiation pressure $p = I_\nu(2/c)\cos\theta$, where I_ν is the intensity of the radiation, during a time interval δt of compression is per unit surface,

$$pv\,\delta t = I_\nu \frac{2v}{c} \cos\theta\,\delta t = (I'_{\nu'} - I_\nu)\,\delta t .$$

It follows that

$$I'_{\nu'} = I_\nu \left(1 + \frac{2v}{c} \cos\theta\right),$$

and hence

$$\frac{I'_{\nu'}}{\nu'} = \frac{I_\nu}{\nu}.$$

But E_ν is proportional to I_ν, which proves the relation quoted on p. 85.

For the particle on p. 96 the Doppler shift formula does not contain the factor 2 because of absorption instead of reflection. But otherwise the situation is the same since the particle is supposed to move slowly in the system K, so that the transition to K' is adiabatic, that is, does not induce Bremsstrahl photons. Therefore the relation $E_\nu/\nu = E'_{\nu'}/\nu'$ of p. 85 also holds here. With the expression

$$\overline{E}_\nu = \frac{c^3}{8\pi\nu^2}\,\varrho_\nu$$

of Sec. 19 (p. 71) the relation quoted on p. 96 readily follows.

[A-8] (p. 106). The formula derived for $\overline{M}$ is called the *Pauli paramagnetism*; see W. Pauli, *Z. Physik* **41**, 81 (1926). The orbital diamagnetism which is due to the Larmor precession of the electrons and has the value $-\frac{1}{3}\overline{M}$ is called the *Landau diamagnetism*; see L. D. Landau, *Z. Physik* **64**, 629 (1930).

In a real solid the electrons move in energy bands and the formula is modified. The diamagnetism of single-band electrons was calculated by R. E. Peierls, *Z. Physik* **80**, 763 (1933). The multiband case is considerably more complicated. For a review see C. P. Enz, in *Semiconductors*, Varenna Summer School 1961, edited by R. A. Smith (Academic Press, New York and London, 1963), p. 458.

Index